MAKE PIGS FLY

I0056925

PRODUCT MANAGER'S
BATHROOM BOOK

-OR-

180 Ways to Impress Colleagues
In Meetings, Emails, And Pitches!

ALI RAKHIMOV

MAKE PIGS FLY

Copyright © 2025 by Ali Rakhimov

Go to www.ali.ink to learn more.

All rights reserved. No part of this publication may be reproduced, distributed, or transmitted in any form or by any means, including photocopying, recording, or other electronic or mechanical methods, without the prior written permission of the publisher, except in the case of brief quotations embodied in critical reviews and certain other noncommercial uses permitted by copyright law.

Paperback ISBN: 979-8-9925416-3-2

Hardback ISBN: 979-8-9925416-4-9

Ebook: 979-8-9925416-5-6

In loving memory of Alexander Wingate.

May you rest in peace, brother.

***

Dedicated to my Wingates, who turned my American adventure from possible to downright unforgettable.

CONTENTS

Preface 1

1. Even a blind hog finds an acorn now and then................... 2

2. Crying Over Spilt Milk 3

3. Drinking Your Bathwater 4

4. All Hat, No Cattle... 5

5. That dog won't hunt.. 6

6. When Pigs Fly 7

7. Jack of All Trades 8

8. Many Ways to Skin a Cat 9

9. Is the juice worth the squeeze? 10

10. Putting lipstick on a pig... 11

11. Ship Without a Harbor 12

12. Any Port in a Storm 13

13. Cinderella Story.. 14

14. Picking Fly Shit from Pepper..................................... 15

15. David and Goliath 16

16. FUBAR.. 17

17. The Golden Rule.. 18

18. Albatross Around Our Neck 20

19. Let the big dog eat 21

20. Run it up the flagpole 22

21. Need to Punt 23

22. Turn a sow's ear into a purse 25

23. Touch all the bases 26

24. Pull the Trigger 27

25. Put on your big boy (or girl) pants 28

26. No Traction 30

27. Face the Music 31

28. Leave Money on the Table 32

29. Nothing ventured, nothing gained 33

30. Put the cart before the horse 34

31. Measure twice, cut once 36

32. A penny saved is a penny earned 37

33. Imitation is the sincerest form of flattery 38

34. A barking dog never bites 39

35. Don't judge a fish by its ability to climb a tree 40

36. Mind your Ps and Qs 42

37. Sell the sizzle, not the steak 43

38. One Foot in the Grave 44

39. Laundry List 45

40. Fighting Fires 46

41. Wiggle Room 48

42. Tighten the Screws 49

43. Sitting on the Fence 50

44. Faster Hors 52

45. Better Mousetrap 53

46. Time-box a decision 54

47. An Uphill Battle 55

48. Pigeon-holing 57

49. Pull the plug 58

50. Dance with those who brought [sic] you 59

51. No Dog in the Fight 61

52. Bake-off 62

53. Blind Spots 63

54. Biases cloud judgments 65

55. Getting in Edgewise 66

56. Staying on Point 67

57. Drawing a Blank 69

58. When the Dust Settles 70

59. Ripple Effect 71

60. Upping the Ante 73

61. Focus on the North Star 74

62. Not everything that glitters is gold 75

63. Testing Scenarios 77

64. Coming down the Homestretch 78

65. Hunky-Dory 79

66. Baby Steps 81

67. Training Wheels 82

68. Rock and Roll 83

69. Slow Jam 85

70. Put Guardrails in Place 86

71. Take a stab at it 87

72. Tip of the iceberg 89

73. Baked in 90

74. Keep you up at night 91

75. Blowing Hot Air 93

76. Going to town 94

77. Scope Creep 95

78. Calm before the Storm 97

79. Touchdown 98

80. Putting your money where your mouth is 99

81. Jump the Gun 101

82. Fact Check 102

83. Nickel and Dime 104

84. Tail off 105

85. Tailwinds 106

86. Headwinds 108

87. Chew on something 109

88. Can't see the forest for the trees 110

89. Throwing out the Baby with the Bathwater 112

90. Fish or Cut Bait 113

91. Swing for the fences 114

92. Hitting the high spots 116

93. Focus on the big picture 117

94. Spending Dollars to Save Dimes 118

95. Getting Lost in the Weeds 120

96. Bells and Whistles 121

97. Drowning in Detail 122

98. Touch Base 124

99. Touchpoints 125

100. Cash Cow 126

101. Sweet Spot 128

102. Hit the ground running 129

103. Best thing since sliced bread 130

104. Early bird gets the worm 132

105. Too good to be true 133

106. Second mouse gets the cheese 134

107. Go Viral 136

108. Devil in the Details 137

109. Rotten Apple 138

110. Eager Beaver 140

111. Deep Dive 141

112. Drain the swamp 142

113. Going Whole Hog 144

114. Strike while the iron is hot 145

115. Whistleblower 146

116. Big Fish in a Small Pond 148

117. Number Cruncher 149

118. Whiz Kid 150

119. Gift of Gab 152

120. He who hesitates is lost 153

121. Beat Around the Bush 154

122. Better late than never 156

123. Cut your losses 157

124. Blessing in Disguise 158

125. Bite the bullet 159

126. Barking Up the Wrong Tree 161

127. A bird in the hand is worth two in a bush 162

128. Paint with a broad brush 164

129. Top Dog 165

130. Stick Out Like a Sore Thumb 166

131. Copycat 168

132. Egg on Your Face 170

133. Lean in 171

134. Best of Both Worlds 172

135. Don't judge a book by its cover 174

136. Getting over your tips (of snow skis) 175

137. Silver Lining 176

138. Other Fish in the Sea 177

139. Bite off more than you can chew 179

140. Break the Ice 180

141. Drive a Hard Bargain 181

142. Bang for the Buck 183

143. Pick someone's Brain 184

144. Play it by ear 185

145. Word of Mouth 187

146. Belt and Suspenders Man 188

147. Let the cat out of the bag 189

148. Chicken or Egg 191

149. Keep an ear to the ground 192

150. Needle in a Haystack 193

151. Bet the Farm 195

152. Spanner in the Works 196

153. Walk a Tightrope 197

154. Spread yourself too thin 199

155. Cut off your nose to spite your face 200

156. Test the Waters 202

157. White Elephant 203

158. Elephant in the Room 204

159. Out of the Blue 206

160. Take the bull by the horns 207

161. A dollar short and a minute late 208

162. Miss the Boat 210

163. Leave it on the field 211

164. The die is cast 212

165. Think outside the box 214

166. Learning Curve 215

167. By the Book 217

168. Cut Corners 218

169. Chop-chop 219

170. Grease the wheels 221

171. Nuts and Bolts 222

172. Run into a Buzzsaw 223

173. Blow a Fuse 225

174. Murphy's Law 226

175. All Sound and Fury 227

176. Blitz 229

177. The short and curlies 230

178. Between a rock and a hard place 231

179. Put your eggs in a single basket 233

180. No Stone Unturned 234

BONUS 236

181. Let sleeping dogs lie 237

182. Beat a Dead Horse 238

183. Hard to Swallow 239

184. Saucered and Blowed 241

185. Screen door in a Submarine 242

186. Chew the Fat 244

187. One-track Mind 245

188. Blank Check 246

PREFACE

"When pigs fly" has long been the punchline for pipe dreams—an offbeat idiom hinting that some things just can't be done. But after countless stories of scrappy go-getters (often immigrants) pushing beyond dead ends—with luck, hustle, and the nudge of supportive mentors—I'm convinced we can all make the "impossible" possible. Hence, **Make Pigs Fly**.

This quirky bathroom book was inspired by a discovery I made as a high school exchange student in 2002: **The Great American Bathroom Book**—a glorious testament to squeezing knowledge into the smallest of "me-times." Now, in an age of infinite digital distraction, we product managers especially owe it to ourselves to reclaim those precious pauses for professional growth.

And if you're squeamish about reading on the throne, don't panic: a 2002 University of Arizona study (gasp!) reassured us that paper isn't exactly a VIP lounge for germs.

It only took me four years of procrastination to wrap this up—but I hope this little volume becomes a trusty companion in your **Product Management** journey and a salute to the relentless drive of immigrants carving out their place in the world.

Enjoy—and may it fire up a spark or two next time you're taking that well-deserved break!

1. EVEN A BLIND HOG FINDS AN ACORN NOW AND THEN

WHAT/WHEN

Always check the data to verify the result.

> "Before we launch this product, we had better check the numbers—after all, even a blind hog finds an acorn now and then, but we don't want to rely on luck!"

AS A PRODUCT MANAGER

Sometimes, success feels like rummaging through chaos and accidentally stumbling upon a hidden gem. "Even a blind hog finds an acorn now and then" perfectly sums up the uncanny reality that, once in a while, even the most clueless among us can strike it lucky. In the business world, this might mean anything from a half-baked app suddenly going viral to a forgotten marketing campaign pulling in unexpected sales.

A memorable example of "Even a blind hog finds an acorn now and then" is when Nintendo launched the Wii back in 2006. They took a gamble on quirky motion controls at a time when no one was sure if flailing around the living room would replace traditional gaming. Lo and behold, they found their acorn, capturing not just hardcore gamers but also your grandma, your pet, and everyone in between—leading to a worldwide phenomenon and record-breaking sales.

For Product Managers, "Even a blind hog finds an acorn now and then" is a reminder that sometimes the best-laid product roadmaps get overshadowed by sheer happenstance. It's that humbling moment when a random feature you almost scrapped is suddenly hailed as a masterstroke by users, while your meticulously planned flagship initiative fizzles out. Ultimately, being ready to pounce on these accidental wins is what keeps the job interesting—because occasionally, luck really is the ultimate growth hack.

2. CRYING OVER SPILT MILK

WHAT/WHEN

Don't dwell on past mistakes or missed opportunities.

> "There is no use crying over spilt milk. The past is over and done.
> We need to move on."

AS A PRODUCT MANAGER

Picture this: you knock over your morning latte. Sure, it's annoying, but once it's all over the floor, you've got two options: cry about it or grab some paper towels. The same goes for product management. If something's gone wrong and there's no time machine handy, lingering on what happened doesn't fix a thing.

Case in point: the infamous iPhone 4 antenna debacle. Apple's product managers faced a chorus of discontented users dropping calls faster than you can drop a mug. They couldn't undo the design flaw, but they rolled out free cases faster than a barista whipping up a latte for a line of caffeine-hungry customers. Problem (mostly) solved.

For Product Managers, "crying over spilt milk" is less about dramatic weeping and more about accepting that you can't change what's already spilt—so you might as well grab a mop and fix what's left. It's juggling disappointment with a semi-forced smile, like discovering your beta sign-up link was broken the entire day while some teammate insists it's the perfect time to add a new feature (spoiler: no, Chad, it's not). Sometimes, it feels like you're the guilty child, the exasperated parent, and the puzzled onlooker all in one. But hey, no sense in drowning your sorrows in spilled milk when you can learn from the puddle, reorder groceries, and move on to the next big launch.

3. DRINKING YOUR BATHWATER

WHAT/WHEN

Don't believe your own hype.

> **"We should have tested the result instead of drinking our own bathwater."**

AS A PRODUCT MANAGER

Ever catch yourself getting way too enamored with your own brilliant product ideas? That's what we call "drinking your own bathwater." (Yes, it sounds gross—and for good reason.) It's when you're so wrapped up in your own epic plans that you fail to notice your team or customers quietly raising their hands, wondering if anyone actually wants that "innovation" you're cooking up.

Amazon's product managers basically threw a party around the Fire Phone concept without checking the guest list first—turns out, no one wanted to attend. Their "revolutionary" features didn't spark joy, and the phone tanked, costing Amazon a fortune. Talk about a reality check.

If you find yourself nodding vigorously at your own pitch—without looking around to see if everyone else is nodding too—it might be time to step away from the tub. Make sure to listen to your team, customers, and stakeholders before you fall in love with your own ideas. Because trust me, drinking your own bathwater never tastes as good as you think.

4. ALL HAT, NO CATTLE

WHAT/WHEN

Consider the source of information.

> **"Joe's promises are worthless. He's all hat and no cattle. He can't deliver."**

This idiom suggests that someone talks a big game but doesn't have the skills or resources to back it up. For product managers, this means that they need to deliver on their promises and have the skills and resources to do so.

A real-world example of this is the development of Google Glass. Google's product managers talked up the device's capabilities, but it didn't live up to the hype. The glasses were expensive, had a short battery life, and didn't offer enough useful features.

Product managers should keep this idiom in mind when promoting their products. They should make sure they have the skills and resources to deliver on their promises and not oversell their products.

5. THAT DOG WON'T HUNT

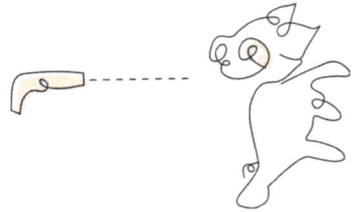

WHAT/WHEN

Abandon bad ideas quickly.

> "It's a mistake to trust that recommendation. That dog won't hunt!"

AS A PRODUCT MANAGER

"That dog won't hunt" is just a colorful way of saying, "Nope, this is never gonna work." Imagine spending hours training your pup to fetch, only to watch it lie there lazily, ignoring every squeaky toy you toss. Sometimes, with certain ideas or projects, you can try, and try, and try… but the sad truth is that dog's just not gonna bring anything back.

In the realm of product management, this idiom reminds us that not every idea is a winner, no matter how cute or promising it seems at first. A golden example (pun intended) is Google Wave. It had all the bells, whistles, and hype—but people simply shrugged, scratched their heads, and went back to their usual tools. Despite oodles of time, energy, and resources, Google eventually had to admit defeat and put Wave out to pasture.

It's never fun admitting your pet project isn't the next big thing, but real talk: if something doesn't gain traction, call it quits. Don't keep tossing treats to a dog that refuses to get off the couch. Cut your losses and pour your heart into the projects that actually have potential to fetch results—because a product that won't hunt is just taking up space on the rug.

6. WHEN PIGS FLY

WHAT/WHEN

When an unlikely or surprising event occurs

"Amazon will be bigger than Walmart when pigs fly."

AS A PRODUCT MANAGER

"When pigs fly"—or in other words, "Yeah, that's never gonna happen." For product managers, this cheeky idiom is a friendly reminder to keep things real. Overpromising on the impossible (like hogs sprouting wings) is a surefire way to bungle your customers' trust.

Need proof? Remember Google Glass—the futuristic headgear that was hyped to turn us all into tech-savvy cyborgs? Everyone was buzzing…until privacy issues, a sky-high price tag, and a few awkward design choices sent this pig crashing to the ground. Google tried to give it another go, but the wings just never grew.

The lesson? Product managers must avoid pie-in-the-sky promises. They should set achievable goals, communicate clearly, and track market signals to pivot as needed—before launching a product that's more fantasy than reality. If an idea's doomed to fail, don't waste time strapping little wings on it. Let it go and focus on the concepts with real potential. After all, real success is much sweeter than bacon-flavored pipe dreams.

7. JACK OF ALL TRADES

WHAT/WHEN

A person with broad knowledge but little specialization.

"Product managers need to be Jacks of all trades."

AS A PRODUCT MANAGER

For product managers, this idiom highlights the importance of having a diverse skill set and being able to work across different functions, but also the need to specialize and become an expert in a particular area.

A real-world example of this is Elon Musk, who is known for his diverse skills and expertise in several areas, including technology, engineering, and business. However, he also understands the importance of specialization, as seen in his leadership role at SpaceX and Tesla, where he focuses on his areas of expertise and delegates other responsibilities to experts in those fields.

Product managers should strive to have a diverse skill set but also recognize the importance of specialization. They should focus on their strengths, work with experts in other areas, and continuously learn and improve their skills.

8. MANY WAYS TO SKIN A CAT

WHAT/WHEN

There is usually more than one way to accomplish a goal.

> **"Competing with Walmart on low price is dangerous. There are many ways to skin a cat. We need to review alternative strategies."**

This idiom refers to the idea that there are multiple ways to achieve a goal. For product managers, this idiom emphasizes the importance of creativity and flexibility in problem-solving.

A real-world example of this is Airbnb, which disrupted the hotel industry by offering a new way of booking accommodation. They found a new approach to address a customer pain point, providing people with a unique and authentic travel experience while saving money. They achieved their goal by thinking outside the box and offering a new solution.

Product managers need to have an open mind and be creative in finding solutions to problems. They should encourage their team to think creatively and experiment with new ideas to achieve their goals.

9. IS THE JUICE WORTH THE SQUEEZE?

WHAT/WHEN

Some goals are not worth the cost and effort to reach.

> "Adding an extra year to the warranty may not add to sales. Is the juice worth the squeeze?"

AS A PRODUCT MANAGER

This idiom refers to the idea of weighing the cost and benefits of a decision or action. For product managers, this idiom highlights the importance of analyzing the return on investment (ROI) and making data-driven decisions.

A real-world example of this is Amazon's decision to acquire Whole Foods. Amazon analyzed the potential ROI of the acquisition and concluded that it would provide them with access to new customers, expand their distribution network, and strengthen their position in the grocery market. The acquisition was a success, and Amazon continues to grow its grocery business.

Product managers should always weigh the cost and benefits of their decisions and make data-driven decisions. They should have a clear understanding of the ROI of their product and continuously monitor its performance to ensure its success.

10. PUTTING LIPSTICK ON A PIG

WHAT/WHEN

Attempting to recover from a marketing disaster.

> "Our product has too many flaws and costs too much. Adding free delivery is just putting lipstick on a pig."

This idiom refers to the act of trying to make something that is inherently unappealing or unsatisfactory look or seem better than it actually is. This can be especially relevant for product managers, who may be tasked with improving a product that has serious flaws or limitations.

One example of "putting lipstick on a pig" in the tech world can be seen in the numerous attempts to improve Microsoft's Internet Explorer web browser. Despite its reputation for being slow, clunky, and outdated, Microsoft continued to release updates and improvements to the browser, attempting to make it more competitive with the likes of Google Chrome and Mozilla Firefox. However, many users still found Internet Explorer to be a subpar browser, and it ultimately fell out of favor in the market.

For product managers, it's important to recognize when a product is fundamentally flawed and cannot be significantly improved through minor updates or cosmetic changes. Rather than wasting time and resources on trying to make a subpar product look better, it may be more productive to invest in developing a completely new and improved product that addresses the issues of the original.

11. SHIP WITHOUT A HARBOR

WHAT/WHEN

Ideas that lack sponsorship.

> **"The proposal to co-market the toys with Disney doesn't have a sponsor. It's a ship without a harbor."**

AS A PRODUCT MANAGER

This phrase refers to the act of launching a product or service without having a clear plan or strategy for how it will succeed in the market. This can be a dangerous approach for product managers, who are responsible for ensuring that their products are successful and profitable.

One real-world example of "shipping without a harbor" can be seen in the case of Google Glass. When the wearable device was first released in 2013, it generated a great deal of hype and excitement, but ultimately failed to gain widespread adoption. This was largely due to a lack of clear use cases and value propositions for the device, which left many potential users unsure of why they would need or want to use it.

For product managers, it's essential to have a clear plan in place for how a product will succeed in the market before launching it. This may involve conducting market research, developing a marketing strategy, and identifying target audiences and value propositions. Without a clear harbor in mind, a product may flounder and fail to gain traction in the market.

12. ANY PORT IN A STORM

WHAT/WHEN

Efforts to recover from a potential disaster.

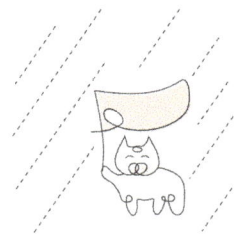

"Online sales are destroying our business. We need to find any port in this storm to survive."

AS A PRODUCT MANAGER

This idiom refers to the act of settling for less than ideal options or solutions when facing difficult or challenging circumstances. In the business world, this can be seen when companies are forced to make compromises or sacrifices in order to stay afloat or meet their goals.

One example of "any port in a storm" in the tech industry can be seen in the case of Apple's decision to partner with Intel for the production of its Macintosh computers in the mid-2000s. While Apple had previously relied on IBM and Motorola for its processors, it was struggling to keep up with the performance improvements being made by Intel. Despite concerns about the compatibility of Intel processors with Apple's software, the company ultimately made the decision to switch to Intel in order to stay competitive.

For product managers, it's important to recognize when compromise is necessary in order to achieve success. This may involve making trade-offs between cost, quality, and time-to-market, or accepting less than ideal solutions in order to meet customer needs or demands.

13. CINDERELLA STORY

WHAT/WHEN

An accomplishment so unexpected it must be a fantasy.

"We have the opportunity to accomplish great things, to be a Cinderella story."

AS A PRODUCT MANAGER

This phrase is often used to describe a situation where an underdog rises from obscurity to achieve great success. For product managers, this is an important concept to keep in mind when developing new products or features.

One real-world example of a Cinderella story is the rise of Zoom, the video conferencing software. Prior to the COVID-19 pandemic, Zoom was a relatively unknown company, with a market value of around $20 billion. However, as remote work became more common, Zoom's user base exploded, and its market value skyrocketed to over $150 billion.

For product managers, the lesson here is that even a small, unknown company can achieve great success with the right product at the right time. By focusing on user needs and delivering a high-quality product, even the smallest companies can compete with the giants in their industry.

14. PICKING FLY SHIT FROM PEPPER

WHAT/WHEN

Considering trivia important

> "We've tested and tested, reviewed all the data time and again. Delaying the start for more study is just picking fly shit out of pepper."

AS A PRODUCT MANAGER

This idiom refers to a situation where someone is nitpicking or obsessing over small details to the point of being counterproductive. This is an important concept for product managers to keep in mind, as it can be easy to get bogged down in minor details and lose sight of the big picture.

One real-world example of picking fly shit from pepper is the development of Apple's Maps app. When the app was first released, it was widely criticized for its inaccuracies and lack of features. However, rather than addressing the major issues with the app, Apple spent a significant amount of time and resources fixing minor details like the placement of certain landmarks.

For product managers, the lesson here is that it's important to prioritize issues based on their impact on the user experience. While it's important to pay attention to detail, focusing too much on minor issues can lead to wasted resources and missed opportunities to address larger problems.

15. DAVID AND GOLIATH

WHAT/WHEN

A situation when a smaller company competes with a dominant rival

> **"Grabbing market share from Apple is a real challenge. We're a David versus Goliath, so we need to be extra creative."**

AS A PRODUCT MANAGER

This idiom refers to a situation where a smaller, weaker opponent is able to defeat a larger, stronger opponent through strategy and determination. For product managers, this is an important concept to keep in mind when competing against larger, more established companies.

One real-world example of a David and Goliath story is the rise of Slack, the team collaboration software. When Slack was first launched, it was up against much larger competitors like Microsoft and Google. However, through a combination of innovative features and a user-friendly interface, Slack was able to carve out a niche in the market and become a major player in the industry.

For product managers, the lesson here is that it's possible to compete with larger companies by focusing on user needs and delivering a high-quality product. By being nimble and responsive to changing market trends, even the smallest companies can succeed against larger competitors.

16. FUBAR

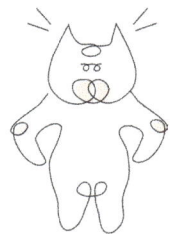

WHAT/WHEN

A situation described as "Fouled Up Beyond All Recognition."

"Why does the App crash when users try to checkout?
This is FUBAR."

AS A PRODUCT MANAGER

This idiom is used to describe a situation that has gone completely awry. For product managers, this is an important concept to keep in mind when managing complex projects and dealing with unexpected challenges.

One real-world example of a FUBAR situation is the development of Boeing's 737 MAX aircraft. Due to a software issue, the aircraft was involved in two fatal crashes, leading to significant regulatory scrutiny and reputational damage for the company.

For product managers, the lesson here is that even the most well-planned projects can go wrong, and it's important to have contingency plans in place to deal with unexpected challenges. By being proactive and transparent in their communication with stakeholders, product managers can mitigate the impact of FUBAR situations and minimize the risk of long-term damage to their company's reputation.

17. THE GOLDEN RULE

WHAT/WHEN

A basic principle that should always be followed to ensure success in general or in a particular activity.

> **"I may greatly disagree with you, but I live by the Golden Rule of respecting everyone's right for an opinion."**

AS A PRODUCT MANAGER

This rule emphasizes the importance of empathy and understanding your customers' needs, as well as your team's needs. By treating your team members with respect and empathy, you build trust and collaboration, resulting in higher team productivity and a better product outcome. This rule also applies to your customers. By understanding their needs, you can create a product that meets their expectations and delivers a positive customer experience.

For example, Apple is known for applying the golden rule to its design philosophy, creating products that are intuitive and easy to use. Apple's founder, Steve Jobs, famously said, "You've got to start with the customer experience and work backward to the technology." This approach has resulted in Apple's success in creating user-friendly products that appeal to a broad customer base.

For Product Managers, the golden rule isn't just a feel-good mantra—it's a survival skill. It's about getting into the trenches with your team, listening, and really *seeing* them as people, not just as roles. It's about knowing when to push and when to step back, about having the guts to say, "What do you think?" instead of always trying to have the answers. And with customers, it's not just about user personas or survey results—it's about getting into their shoes, feeling their frustrations, and obsessing over how to solve their problems. The golden rule isn't fluff; it's how you build products that people love and teams that will fight like hell to make them succeed. It's not easy, but it's what separates good Product Managers from great ones.

18. ALBATROSS AROUND OUR NECK

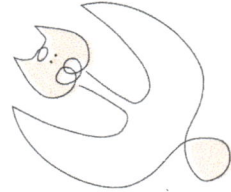

WHAT/WHEN

A marketing or strategic disadvantage that cannot be controlled.

> **"The lack of our product's customization is an albatross around our neck."**

AS A PRODUCT MANAGER

This idiom refers to a heavy burden or problem that weighs us down. As a product manager, it's essential to identify any potential problems or challenges and address them early on to avoid them becoming an albatross around your team's neck. This includes understanding potential market risks, ensuring that product features align with customer needs, and addressing any technical issues that may arise.

For example, the Boeing 737 Max airplane was grounded globally after two fatal crashes, and investigations revealed significant design flaws in the aircraft. This incident was a severe albatross around Boeing's neck, and the company's reputation was severely damaged. Product managers must ensure that they thoroughly vet all aspects of their products, from design to production to avoid any significant issues that could harm their company's reputation.

For Product Managers, the albatross around your neck is more than just a cliché—it's the looming nightmare that could derail your entire product and trash your reputation. You can't shrug off potential issues or sugarcoat your challenges. You have to dig deep, question every assumption, and stay vigilant about what could go wrong. Because once that albatross latches on, it's a relentless weight you'll be forced to carry long after the damage is done.

19. LET THE BIG DOG EAT

WHAT/WHEN

Exploiting a marketing/size advantage.

> **"We've got the lowest price product in our market. That's a big advantage. It's time to let the Big Dog eat."**

AS A PRODUCT MANAGER

This idiom is often used in the business world to encourage aggressive, bold actions that can lead to significant gains. As a product manager, it's essential to take calculated risks and make strategic decisions that can lead to big wins for your company. This may include investing in new technologies, entering new markets, or developing innovative products that disrupt the industry.

For example, Tesla's CEO, Elon Musk, has famously encouraged his team to "let the big dog eat" when it comes to innovation. Tesla has disrupted the automotive industry with its electric cars, and they continue to invest in new technologies such as self-driving cars and solar power. Product managers must be willing to take risks and think outside the box to create products that can drive significant growth for their companies.

For Product Managers, "letting the big dog eat" isn't just some cliché—it's a rallying cry to lean in hard when the stakes are highest. It's about throwing your weight behind bold ideas that, yes, might fail spectacularly, but could also blow open entirely new markets. You don't want to wait on the sidelines, hoping something miraculous happens; you want to be the one making it happen. The truth is, if you keep muzzling the big dog, you'll never know what game-changing opportunities could've been yours for the taking.

20. RUN IT UP THE FLAGPOLE

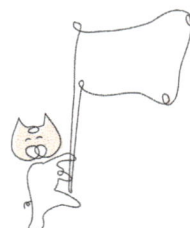

WHAT/WHEN

Testing a new idea or product for reaction.

> **"That sounds like a good idea. Let's run it up the flagpole and see what others think."**

AS A PRODUCT MANAGER

This idiom refers to the process of presenting an idea or proposal to a higher authority for approval. As a product manager, it's crucial to get buy-in from stakeholders, including executives, investors, and team members, to ensure that your product is aligned with your company's goals and objectives. This may involve presenting your product roadmap, market research, and financial projections to get approval and secure funding.

For example, Airbnb's product managers had to "run it up the flagpole" when they first pitched their idea of allowing homeowners to rent out their homes to travelers. They had to convince investors that their platform could disrupt the traditional hotel industry and create a new market for home-sharing. Today, Airbnb is a multi-billion dollar company with a global presence, thanks to the product managers who successfully ran their idea up the flagpole.

For Product Managers, "running it up the flagpole" isn't just some corporate buzzword you throw around in meetings—it's a real, gritty process of testing your conviction in front of people who can make or break your idea. You have to stand behind your roadmap, show why it's worth betting on, and be willing to face tough questions if it doesn't fly. It's not for the timid, but if you can own your pitch and lead with passion, you'll find out fast who salutes—and who doesn't. That's how you earn the resources and support to build something extraordinary.

21. NEED TO PUNT

WHAT/WHEN

Recognizing a loss with no hope of recovery.

> "We've redesigned the product, lowered its price, and developed a new marketing campaign. Everything we've tried has failed. It's time to punt and move on to the next project."

AS A PRODUCT MANAGER

This phrase refers to the situation when a product manager faces a difficult decision or problem that they cannot solve or address. In such cases, the product manager may need to take a step back, reassess the situation, and consider a different approach. Sometimes, the best course of action may be to defer the decision or problem to someone else, such as a higher-level manager or a different department within the organization.

One real-world example of when a product manager may need to punt is when faced with a major strategic decision that requires input from multiple stakeholders. In such cases, the product manager may need to seek guidance from senior management, board members, or other experts in the field before making a final decision.

For Product Managers, the "need to punt" can feel like admitting defeat, but in reality, it's about having the self-awareness to know when you're out of your depth. It's a raw moment of honesty where you consciously hand off the ball to someone with better insight, broader context, or greater influence. Instead of stubbornly pushing forward and risking the product's success, you make the tough call to pivot, collaborate, or escalate. By punting at the right time, you keep the bigger vision in play—and that's what ultimately drives a product forward.

22. TURN A SOW'S EAR INTO A PURSE

WHAT/WHEN

Making the best of a bad situation.

> **"The product doesn't meet our expectations but is too expensive to abandon. Our charge is to turn a sow's ear into a purse."**

AS A PRODUCT MANAGER

This idiom refers to the act of transforming something that is perceived as low-quality or unattractive into something valuable or desirable. In the context of product management, this idiom can be applied to situations where a product manager is tasked with improving a product that has underperformed or failed to meet customer expectations.

One real-world example of turning a sow's ear into a purse in the tech industry is the transformation of Apple's iPod into the iPhone. When Apple first introduced the iPod in 2001, it was a popular music player but lacked many of the features that consumers expected from a mobile device. However, Apple was able to turn the sow's ear of the iPod into the purse of the iPhone by adding phone and internet capabilities to the device, making it a game-changer in the mobile phone market.

For Product Managers, the essence of turning a sow's ear into a purse lies in seeing the potential where others see problems. It's not just about fixing what's broken; it's about reimagining what's possible. Whether it's repurposing an underwhelming feature into a core product strength or identifying an underserved market that aligns with an existing asset, the transformation requires vision, grit, and relentless curiosity. It's in the unglamorous, raw moments of questioning assumptions and pushing boundaries that the true magic happens. After all, every purse started as a sow's ear in someone's eyes.

23. TOUCH ALL THE BASES

WHAT/WHEN

Being fully prepared with the right approvals.

> **"Before we spend any time on this project, I need to touch all the bases to see if there is support for it."**

AS A PRODUCT MANAGER

This idiom refers to the act of ensuring that all necessary steps have been taken to achieve a goal or complete a task. In the context of product management, this idiom can be applied to situations where a product manager is responsible for launching a new product or feature and must ensure that all aspects of the launch are addressed.

One real-world example of touching all the bases in the tech industry is the launch of a new software application. To ensure a successful launch, the product manager must touch all the bases by conducting market research, identifying customer needs, designing the user interface, developing the software, testing the software for bugs and usability issues, and creating a marketing and distribution plan.

For Product Managers, touching all the bases means more than just checking off tasks on a to-do list—it's about truly understanding and owning every part of the process. It's knowing that if you miss even one base, the entire play could fall apart. Maybe it's a missed edge case in QA, a gap in communication with stakeholders, or an overlooked user pain point that could tank the experience. The raw truth is, touching all the bases is messy, imperfect work—it's late-night whiteboarding sessions, hard conversations about priorities, and sometimes pivoting when you thought you were done. But that's where the magic happens, in those gritty moments of chasing down every detail to create something that truly lands.

24. PULL THE TRIGGER

WHAT/WHEN

Decide, take action, and give the go-ahead.

> "We have done the research, checked the data, and discussed the project from end to end. It's time to pull the trigger."

This idiom is often used in the context of decision-making. It refers to the moment when a person or a team decides to move forward with a particular course of action, often after careful consideration and analysis of the potential risks and benefits. For product managers, the ability to pull the trigger at the right time can be crucial for the success of a product.

In the tech industry, a good example of a company that pulled the trigger at the right time is Airbnb. In 2013, the company was struggling to gain traction in the market, and its founders were considering shutting it down. However, they decided to take a risk and invest heavily in growth, focusing on improving the user experience and expanding the platform. This decision paid off, and today Airbnb is a global success story.

On the other hand, failing to pull the trigger can lead to missed opportunities and stagnation. Kodak is a classic example of a company that failed to pull the trigger when digital photography was emerging as a disruptive technology. Instead of embracing the change and investing in digital cameras, Kodak clung to its traditional film business, eventually leading to its downfall.

25. PUT ON YOUR BIG BOY (OR GIRL) PANTS

WHAT/WHEN

Take responsibility in tough situations.

> **"I know the rollout was not what we expected, but it was our baby. It's time to put on your big boy pants and accept the consequences."**

AS A PRODUCT MANAGER

This idiom is often used to encourage someone to be more responsible, assertive, or mature. For product managers, this means taking ownership of their projects and being willing to make tough decisions when necessary.

One example of a company that put on its big boy pants is Netflix. In the early days, Netflix was primarily a DVD-by-mail service, but the company recognized the potential of online streaming and began investing heavily in that technology. This decision required a significant shift in the company's business model, but it ultimately paid off, and today Netflix is one of the world's largest streaming platforms.

In contrast, failing to put on your big boy (or girl) pants can lead to missed opportunities and stagnation. Blockbuster, once the dominant player in the video rental industry, failed to adapt to the changing market and the rise of online streaming. Instead of investing in technology and embracing the new model, Blockbuster clung to its traditional brick-and-mortar stores, eventually leading to its demise.

26. NO TRACTION

An idea or campaign that fails to catch on with prospective customers.

> **"I thought Gerber Singles was a winner, but it never gained traction with the public."**

AS A PRODUCT MANAGER

This idiom refers to a situation where a product or a project is failing to gain momentum or make progress. For product managers, this can be a warning sign that something needs to change in order to get the project back on track.

A good example of a company that faced "no traction" but was able to turn things around is Slack. When Slack was first launched in 2013, it struggled to gain users, and many people dismissed it as just another chat app. However, the company continued to iterate on the product, adding features and improving the user experience. Today, Slack is a hugely successful communication platform used by millions of people around the world.

For Product Managers, "no traction" is a blunt reality check. It's the point where all the effort, planning, and resources feel like they're vanishing into thin air without making an impact. It's frustrating, humbling, and, frankly, a little embarrassing. But it's also the moment where you have to confront the uncomfortable questions: Are we solving a real problem? Are we even talking to the right people? It's not about spinning the story or doubling down on what's not working—it's about peeling back the layers, understanding what's missing, and finding a way to genuinely connect with the people who need what you're building.

27. FACE THE MUSIC

WHAT/WHEN

Accept bad consequences without alibis or excuses.

> **"My department overestimated the appeal of the ad message. I have no excuses, and I'm prepared to face the music."**

AS A PRODUCT MANAGER

This idiom means to confront the unpleasant consequences of one's actions or decisions. For product managers, facing the music is crucial as it allows them to take responsibility for their actions and learn from their mistakes. This is especially important in the tech industry where product launches can be risky and the consequences of failure can be significant.

For example, in 2016, Samsung faced a major crisis with their Galaxy Note 7 smartphones catching fire. Instead of denying the problem, Samsung faced the music by recalling the phones and taking responsibility for the issue. This allowed them to learn from the problem and improve their products in the future.

For Product Managers, "facing the music" is more than just a professional obligation—it's a moment of truth. It's standing in the eye of the storm when a launch flops, a bug slips through, or user adoption tanks. It's about owning the fallout and acknowledging, "This happened on my watch." It's raw, uncomfortable, and often public, but it's also where growth begins. A great PM doesn't just endure these moments—they use them to fuel better decisions, build stronger teams, and ultimately deliver products that resonate. Because in the end, the music you face today shapes the harmony you create tomorrow.

28. LEAVE MONEY ON THE TABLE

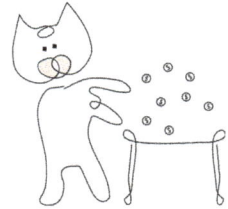

WHAT/WHEN

Failure to capture the full benefit of a superior position.

> **"I negotiated the price for 12 weeks of a full-page Sunday advertisement with the Times. Even though I got a price lower than I expected, I think I left money on the table."**

AS A PRODUCT MANAGER

This idiom means to miss out on an opportunity to maximize profits. For product managers, leaving money on the table can be detrimental to the success of their products and companies. It is important for product managers to ensure that they are monetizing their products effectively and not missing out on potential revenue streams.

For example, when Apple first launched the iPhone, they only offered it through AT&T in the United States. However, they later expanded their carrier partnerships to include other major carriers such as Verizon and Sprint, allowing them to reach a wider audience and increase their profits.

For Product Managers, leaving money on the table is more than just a lost opportunity—it's a failure to truly understand the value your product brings and how people interact with it. It's often the result of not asking the hard questions: Are we charging what this is worth? Are we making it accessible to everyone who wants it? Are we solving the right problems for the right audience? It's not about squeezing every dollar out of customers; it's about respecting the effort and resources poured into building something worthwhile and making sure it's reaching its full potential. Every missed dollar reflects a decision—or lack of one—that could have been better.

29. NOTHING VENTURED, NOTHING GAINED

WHAT/WHEN

Sometimes taking a risk is necessary to reach a goal.

> "I believe in this campaign. I know it will work, so I put my reputation behind it to get approval. After all, nothing ventured, nothing gained."

This phrase means that without taking risks, one cannot expect to achieve success or gain rewards. For product managers, taking calculated risks is essential to innovation and growth. In the tech industry, companies that take risks and innovate are often the most successful.

For example, when Google launched its self-driving car project, it was a risky move as it was a new and untested technology. However, by taking this risk and investing in the project, Google became a leader in the autonomous vehicle industry.

Another example is the launch of Airbnb. When the founders of Airbnb first started their company, they faced many challenges and risks such as regulatory hurdles and concerns over safety. However, by taking these risks and innovating in the sharing economy space, Airbnb has become a billion-dollar company.

For Product Managers, the lesson is simple: without risk, there's no reward. True innovation comes from stepping into the unknown with purpose. Setbacks are possible, but so are breakthroughs that redefine markets. Playing it safe may protect you, but it rarely leads to greatness.

30. PUT THE CART BEFORE THE HORSE

WHAT/WHEN

Acting before preparation is complete.

> **"We paid for an upgrade in our service without knowing what we needed. Talk about putting the cart before the horse."**

AS A PRODUCT MANAGER

This idiom means doing things in the wrong order or starting a project before having a clear understanding of the steps required to achieve the desired outcome. This can result in wasted resources, delayed timelines, and unsatisfactory outcomes.

For example, if a tech startup launches a new product without conducting adequate market research or validating the product-market fit, it may result in low adoption rates, negative reviews, and ultimately lead to failure. Similarly, if a company invests in advertising campaigns before optimizing its website's user experience, it may not result in the desired conversion rates.

For Product Managers, "putting the cart before the horse" is that moment you rush into releasing flashy features or pumping money into marketing before you've nailed down the basics—like validating your user needs or creating a solid product roadmap. It might feel exciting at first, but it's a fast track to wasted budgets, disillusioned teams, and a product no one really wants. Think of it like trying to race off in a car with no engine; it might look good on the outside, but you'll go nowhere fast. So slow down, set the right priorities, and let the horse lead the way—because no one wants to watch a brilliant idea collapse under its own misguided haste.

31. MEASURE TWICE, CUT ONCE

WHAT/WHEN

Confirm the data before acting.

> "The projected results seem too good to be true. Let's check the data one more time before recommending the changes in our ads. It's smarter to measure twice and cut once than fix an avoidable mistake."

AS A PRODUCT MANAGER

This phrase emphasizes the importance of careful planning and attention to detail. As a product manager, it's crucial to ensure that you're making informed decisions based on accurate data and insights. Rushing into decisions without adequate preparation can result in costly mistakes.

For example, if a software company releases a new version without thoroughly testing it, it may result in bugs and user dissatisfaction. Similarly, if a company invests in a marketing campaign without measuring its effectiveness, it may not result in the desired ROI.

Measure twice, cut once" isn't just a handy adage for carpenters—it's a survival mantra for Product Managers. It's about taking the time to validate assumptions, ask the tough questions, and make sure the foundation is solid before you move forward. Skipping this step might feel faster in the moment, but it often leads to scrambling later to patch up holes that could have been avoided. Whether it's launching a new feature or pivoting on strategy, every rushed cut comes with the risk of irreversible damage. In product management, precision isn't a luxury; it's the difference between building something users love and something that falls flat.

32. A PENNY SAVED IS A PENNY EARNED

WHAT/WHEN

The amount of investment doesn't dictate its return.

> **"Do we need to spend this amount of money on the rollout? We have little left to make any needed adjustments."**

AS A PRODUCT MANAGER

This idiom emphasizes the importance of financial prudence and discipline. As a product manager, it's crucial to manage resources effectively and optimize costs without compromising quality.

For example, if a tech startup spends a significant amount of money on unnecessary office space or equipment, it may result in a shortage of funds for critical activities like product development or marketing. Similarly, if a company invests in expensive tools without considering their value, it may not result in the desired ROI.

For Product Managers, a penny saved is a penny earned means more than just pinching pennies—it's about having the courage to question every dollar spent and the awareness to channel savings back into the product's lifeblood. Sacrificing what's truly unnecessary can buy you room to innovate, pivot, and fail fast without tanking your budget. At the end of the day, mindful spending isn't about being stingy; it's about protecting your team's creative freedom and building a more resilient path to success.

33. IMITATION IS THE SINCEREST FORM OF FLATTERY

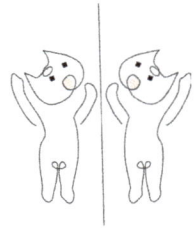

WHAT/WHEN

Copying can be an effective offensive or defensive response to competition.

"Customers buy products that give them the greatest value for their money. XYZ company has raised their warranty term to three years. We should do the same"

AS A PRODUCT MANAGER

This idiom suggests that they should pay attention to what their competitors are doing and learn from their successes and failures. While copying a competitor's product directly is not advisable, understanding what they are doing right and adapting those ideas to fit their own product can be beneficial.

For example, when Apple introduced the iPhone in 2007, it revolutionized the smartphone industry. Many companies quickly realized the potential of smartphones and began creating their own versions. However, some companies simply copied the iPhone's design and features, rather than innovating on their own. These companies failed to achieve the same level of success as Apple. On the other hand, companies like Samsung studied the iPhone's success and made their own improvements, such as larger screens and better cameras, resulting in their own successful product lines. By imitating the success of others and building on it, product managers can create products that stand out in the market.

For Product Managers, the trick isn't to copy but to borrow brilliance and make it your own. Take what works, give it a glow-up, and serve it with your own flavor. Innovation isn't stealing the recipe—it's remixing it into something so good everyone forgets the original.

34. A BARKING DOG NEVER BITES

WHAT/WHEN

Threats and claims of superiority are often empty.

"ABC Company claims that our electric toothbrush is just an inferior knockoff of their product. I guess they haven't read the latest Consumer Reports for both products."

AS A PRODUCT MANAGER

This idiom suggests that people who make the most noise are often the least threatening. For product managers, this means that they should not be overly concerned with competitors who talk a big game but don't have the ability to back it up. Instead, they should focus on their own product and its strengths.

For example, when Google launched its social networking platform, Google+, in 2011, it was touted as a "Facebook killer." However, despite its initial hype, Google+ failed to gain traction and was eventually shut down. Meanwhile, Facebook continued to dominate the social media market.

For Product Managers, the takeaway is this: don't lose sleep over the loud, showy competitors. A barking dog never bites—it just wants attention. Let them bark. Meanwhile, you're the one training your team, refining your product, and delivering real results. Because at the end of the day, users don't care about the noise—they care about the treats, and you're the one handing those out.

35. DON'T JUDGE A FISH BY ITS ABILITY TO CLIMB A TREE

WHAT/WHEN

People are different, with diverse skills, personalities, and interests.

> **"Mike is exceptionally creative but is easily distracted. Bob is a head's down, practical thinker who lacks imagination. By working together, they learn new skills."**

AS A PRODUCT MANAGER

This phrase emphasizes the importance of recognizing individual strengths and weaknesses. For product managers, this means understanding the unique capabilities and limitations of their team members and assigning tasks accordingly.

For example, if a product manager is leading a team to develop a new software product, they may have team members with different strengths and skill sets. One team member may excel at coding, while another may be a strong communicator and presenter. Understanding these individual strengths and assigning tasks accordingly can lead to a more effective and efficient product development process.

For Product Managers, "don't judge a fish by its ability to climb a tree" isn't just a cute phrase—it's a wake-up call to stop trying to force everyone into the same cookie-cutter role. Some team members will thrive when diving deep into code, while others will swim circles around communication or customer research. Your job is to find that sweet spot for each person so they can do what they do best—because a fish stuck on a branch is bound to fail, but give it water and it'll surprise you every time.

36. MIND YOUR PS AND QS

WHAT/WHEN

Focus on your task. Be meticulous and precise.

> **"Before criticizing advertising, mind your Ps and Qs. You do your job and let them do the same. We will have time to share ideas and differences as a group."**

AS A PRODUCT MANAGER

This idiom means to be careful with your behavior, speech, and manners in a particular situation. In the world of product management, this idiom carries significant importance, as the behavior and actions of product managers can have a direct impact on the success or failure of a product.

One way that product managers can "mind their Ps and Qs" is by being diligent in their communication with stakeholders. This includes keeping everyone informed about the progress of the product, setting expectations appropriately, and avoiding making promises that cannot be kept. In the tech industry, a prime example of a product manager who successfully "minded their Ps and Qs" is Sundar Pichai, CEO of Google. Throughout his tenure, he has been known for his calm and thoughtful communication style, which has helped him to navigate complex and often controversial issues with ease. Additionally, he has prioritized employee well-being, introducing policies like extended parental leave and mental health support.

For Product Managers, it's all about keeping it together—saying the right things, not overpromising, and definitely not emailing at 2 a.m. with 'great ideas.' A little tact and a lot of clarity can save your team from chaos and your product from becoming the next cautionary tale.

37. SELL THE SIZZLE, NOT THE STEAK

WHAT/WHEN

Promote the outcome, not the details.

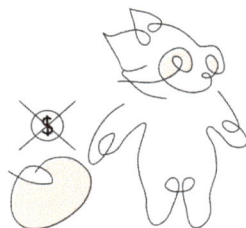

> **"Customers buy products and services for the way they make them feel. Facts are important, but feelings drive the decision. The sizzle sells the steak."**

AS A PRODUCT MANAGER

This phrase emphasizes the importance of highlighting the benefits of a product rather than just its features. In the tech industry, companies often focus on the technical specifications and features of their products, assuming that customers will be impressed by the bells and whistles. However, what really captures the attention of potential customers is how a product can solve their problems and make their lives easier.

A great example of this is Apple's marketing strategy for the iPhone. While the technical features of the device are impressive, Apple's marketing emphasizes the ways in which the iPhone can make users' lives better, such as by providing easy access to social media, high-quality photography, and seamless integration with other Apple products.

For Product Mangers, this mindset is a game-changer. It's not just about listing what your product *does*; it's about painting a vivid picture of how it makes someone's life better, easier, or more exciting. Your customer doesn't care about the gigabytes or algorithms—they care about capturing memories with friends, cutting through daily chaos with efficiency, or feeling like they're ahead of the curve. Speak to the heart, not the hardware. The "sizzle" is the promise, the story, the emotional resonance that makes your product unforgettable. As a Product Manager, your job is to translate features into transformative experiences and make your audience feel, "I need this." That's what sells.

38. ONE FOOT IN THE GRAVE

WHAT/WHEN

The end of a product's life, the beginning of a lingering decline.

> **"The introduction of mobile phones left the pay phone industry with one foot in the grave. It disappeared with the popularity of inexpensive cell phones."**

AS A PRODUCT MANAGER

This phrase is used to describe someone who is very old or sick and is close to death. In the tech industry, this idiom can refer to products or companies that are outdated and struggling to stay relevant.

In order to stay competitive in the fast-paced tech industry, product managers must constantly innovate and adapt to changing market conditions. Products that are slow to evolve or fail to meet changing customer needs can quickly become irrelevant and outdated. One example of a company that struggled with this is Blockbuster. The video rental giant failed to adapt to the rise of digital streaming services like Netflix, and eventually went bankrupt in 2010.

For Product Managers, keeping a product alive isn't about slapping on a fresh coat of paint and hoping no one notices the cracks—it's about knowing when your ship is taking on water and having the guts to build a speedboat instead. Let's face it, in tech, yesterday's shiny idea can become today's punchline faster than you can say 'blockchain.' A product that's barely hanging on doesn't just fade away—it drags everyone down with it, like the office fridge smell no one will admit is their leftover fish. The truth hurts: if you're not evolving, you're embarrassing yourself."

39. LAUNDRY LIST

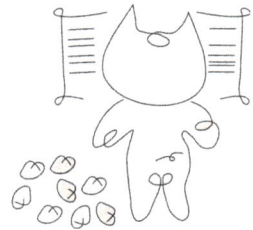

WHAT/WHEN

A set of necessary or suggested process steps, resources, and approval accompanying a specific objective.

> "Before we approve the marketing campaign, we need to complete our laundry list of requirements."

AS A PRODUCT MANAGER

This phrase refers to a long and often overwhelming list of tasks or items that need to be completed. In the tech industry, product managers can be faced with a never-ending laundry list of features, bugs, and other tasks that need to be addressed. In order to manage these competing priorities, product managers need to be able to prioritize tasks based on their importance and impact on the product.

One example of a company that excels at this is Google, which uses a framework called OKRs (Objectives and Key Results) to set priorities and measure progress. By setting clear objectives and focusing on the key results that will have the biggest impact on the product, Google is able to prioritize its laundry list of tasks and ensure that the most important items are addressed first.

For Product Managers, the endless to-do list can feel like being handed a pile of dirty clothes, half of which might not even be yours. You've got to figure out which items are worth cleaning, which ones can wait, and which are just someone else's socks you shouldn't even be dealing with. Prioritizing is less about getting everything done and more about avoiding the temptation to shove the whole mess under the bed and hope no one notices.

40. FIGHTING FIRES

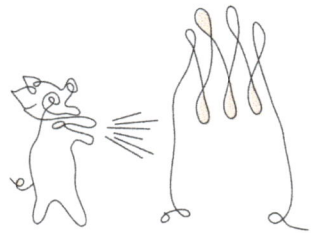

WHAT/WHEN

Constant interruptions to correct avoidable or inconsequential mistakes.

> **"For every step forward, I take two steps back. I'm constantly fighting fires instead of completing my objective."**

AS A PRODUCT MANAGER

It's important to be able to address problems as they arise, especially when they're urgent. This idiom perfectly encapsulates this need to be able to react quickly and efficiently to issues that arise. In the tech industry, this could mean anything from addressing a sudden security breach to fixing a critical bug in a product that's about to launch.

One real-world example of a product manager "Fighting Fires" is when Apple had to deal with the infamous "Antennagate" scandal in 2010. The iPhone 4 had a design flaw that caused signal loss when the phone was held a certain way. The product manager had to quickly respond to the issue and come up with a solution, which ended up being a free case for all iPhone 4 owners.

For Product Managers, "fighting fires" is less about heroism and more about not letting the whole building burn down while you're figuring out why the toaster exploded in the first place. It's juggling chaos with a straight face—like responding to a critical bug that just surfaced hours before launch while someone on the team is asking if they can add one last feature (spoiler: no, Chad, we can't). Sometimes, it feels like you're the firefighter, the arsonist, and the person who forgot to check the smoke detector all at once. But hey, nothing bonds a team like the smell of singed priorities and the adrenaline of a good ol' crisis.

41. WIGGLE ROOM

WHAT/WHEN

A margin of safety, freedom to act in alternative ways.

> **"I'm relatively confident we can deliver the product as designed, but I need some wiggle room on the budget."**

AS A PRODUCT MANAGER

It's important to have a little bit of leeway in your plans, especially when your project scope seems to be changing by the hour. This idiom perfectly captures the value of not boxing yourself into one rigid path when the future might throw a wrench—or three—into the works. In the tech industry, this could mean building an extra week into the development cycle or setting aside some budget "just in case," so you're not in a full-blown panic at the first unexpected hurdle.

A prime example of leveraging "wiggle room" is when Netflix began shifting from mailing DVDs to streaming services. Rather than going all-in on DVDs, they maintained enough flexibility to pivot as broadband speeds improved and user habits changed. That freedom to maneuver took them from a postage-stamp-slinging curiosity to an entertainment giant practically running half your TV screen time.

For Product Managers, "wiggle room" is the secret sauce that prevents your entire timeline from imploding the moment Chad from marketing shows up with his "We absolutely need this feature yesterday!" request (spoiler: no, Chad, we can't). It's the buffer between you and total chaos, letting you handle curveballs without becoming the office stress-ball. Sure, you might feel like a human Stretch Armstrong being yanked in 14 different directions, but at least you won't snap in two when things inevitably get…well, wiggly.

42. TIGHTEN THE SCREWS

WHAT/WHEN

Applying pressure on a process, an outcome, or an individual.

> "We've got to tighten the screws so that the prototype testing phase will happen by the end of February."

AS A PRODUCT MANAGER

"It's important to recognize when a little extra pressure is needed to get things moving in the right direction. This idiom captures the essence of stepping up accountability, whether it's tightening a slack process or pushing a project forward with a bit more urgency. In the corporate world, this can mean anything from nudging a team to meet a looming deadline to cracking down on quality control issues before they spiral.

A classic example of someone "tightening the screws" is Elon Musk during Tesla's infamous "production hell" phase for the Model 3. Facing massive delays and potential financial disaster, Musk literally moved his desk to the factory floor, slept in the building, and put immense pressure on the team to hit production targets. While it wasn't exactly a picnic for anyone involved, it ultimately paid off in turning things around.

For Product Managers, "tightening the screws" is less about being a tyrant and more about finding creative ways to remind everyone that deadlines aren't just theoretical concepts. It's rallying the team to get things done without making them feel like they're on a sinking ship. Sometimes that means a pep talk, sometimes it's a strongly-worded Slack message, and sometimes it's eating the last donut in the break room to fuel your motivational speech about urgency. Just remember, there's a fine line between tightening screws and stripping them—and nobody wants to deal with *that* metaphorical hardware store visit."

43. SITTING ON THE FENCE

WHAT/WHEN

The state of indecision.

> "The market views our computers like those of our competitors. To get them off the fence, we must show them a clear reason for them to pick us."

AS A PRODUCT MANAGER

It's important to make decisions when needed, especially when others are looking to you for guidance. This idiom perfectly encapsulates the hesitation that can arise when faced with difficult choices or competing priorities. In the tech industry, this could mean anything from choosing between two competing feature ideas to deciding how to handle an underperforming team member.

One real-world example of a product manager "Sitting on the Fence" is the debacle around Google's decision-making on messaging apps. For years, Google seemed to oscillate between various platforms—Hangouts, Allo, Duo, Chat, and more—without ever committing fully to one. The lack of a clear direction left users confused and the competition (hello, WhatsApp) free to dominate the market. If "Sitting on the Fence" had a tech mascot, it might just be a Google PM flipping a coin.

For Product Managers, "sitting on the fence" is less about thoughtful deliberation and more about waiting until the fence starts to splinter under the weight of indecision. It's wrestling with FOMO and analysis paralysis simultaneously—like debating whether to pivot or persevere while the sales team is demanding a roadmap update and the dev team is Googling "careers in pottery." Sometimes, it feels like you're auditioning for a role in the circus: the tightrope walker who doesn't know whether to step forward, step back, or just hang out awkwardly in the middle until someone else makes the call. But hey, at least the view from up there gives you time to think...a lot.

44. FASTER HORS

A need to reframe the objective.

> **"For centuries, people thought to breed faster horses to cut travel times. Henry Ford imagined a replacement for the horse that changed civilization."**

AS A PRODUCT MANAGER

It's crucial to understand what people really need versus what they think they need, especially when innovating. The idiom "faster horse" perfectly captures this challenge of balancing customer feedback with visionary thinking. In product development, it's the difference between solving immediate problems and building solutions that customers didn't even know they wanted.

A classic example of avoiding the "faster horse" trap is Henry Ford and the Model T. Supposedly, when asked about customer input, Ford quipped, "If I had asked people what they wanted, they would have said faster horses." Instead of improving horse-drawn carriages, he revolutionized transportation entirely by introducing affordable automobiles, forever changing how people traveled.

For Product Managers, navigating "faster horse" moments is like deciphering cryptic riddles from users who just want a "better version of this thing I already have." It's smiling politely while someone insists the solution to their problem is adding *another* button. It's knowing when to listen and when to nod sagely and say, "What if we just built you a car instead?" After all, innovation isn't about galloping faster—it's about finding a whole new road. And sometimes, that road leads to a world where horses finally get to retire.

45. BETTER MOUSETRAP

WHAT/WHEN

More efficient, effective solutions to common problems.

> "Customers always complain about opening our sealed plastic packaging. We need an easier opening secure package."

AS A PRODUCT MANAGER

It's important to constantly improve and innovate, especially in competitive fields where standing still means falling behind. This idiom encapsulates the drive to create something better than what already exists, to revolutionize rather than settle for the status quo. In business, this could mean redesigning a product that's already successful or rethinking processes that seem "good enough" to make them truly exceptional.

A real-world example of building a "Better Mousetrap" is when Tesla entered the automotive market. Cars already existed, and electric cars weren't a new concept, but Tesla made them cool, high-performing, and aspirational. They didn't just create another car; they created a whole new mousetrap that forced traditional automakers to rethink their designs, strategies, and priorities.

For Product Managers, building a better mousetrap is less about outsmarting rodents and more about outsmarting competitors (although both involve cheese at some level). It's finding the gaps in what people didn't even know they needed and filling them with features no one can live without—like reinventing email so it sorts itself or designing a coffee mug that reminds you to actually drink your coffee while it's still hot. The challenge is in knowing when to stop. After all, no one asked for a mousetrap that syncs with Alexa, keeps a diary of every mouse it catches, and alerts your smartwatch to "celebrate the catch." Sometimes, better is simply smarter, not more complicated.

46. TIME-BOX A DECISION

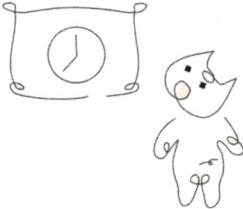

WHAT/WHEN

Putting a deadline on a decision.

> **"If we are to meet the schedule, we need to time-box staffing the team members by March 15."**

AS A PRODUCT MANAGER

It's crucial to make decisions in a timely manner, especially when there are countless priorities vying for your attention. "Timebox a decision" embodies this need to set limits on deliberation and move forward with clarity. In any industry, this could mean anything from choosing a new product direction to resolving a supply chain hiccup before it spirals.

One well-known example of "timeboxing a decision" is when Domino's Pizza decided to overhaul their pizza recipe in 2009. After years of customer complaints about the taste and quality, they set a strict timeline to reinvent their core product. Within months, they developed a new recipe, tested it, and rolled it out nationwide. The time-bound decision to act decisively on customer feedback transformed the company and reinvigorated their brand.

For Product Managers, "timeboxing a decision" is less about rushing and more about preventing endless waffling. It's saying, "We're picking an option by Thursday, even if Kyle in marketing hasn't finished his 15th focus group." It's the discipline of keeping a project moving forward, even if it feels like you're choosing between "bad" and "slightly less bad" options. Sure, there's risk, but as long as you don't timebox yourself into deciding your flagship app should have a pizza tracker that plays opera, you'll probably come out ahead.

47. AN UPHILL BATTLE

WHAT/WHEN

Persevere despite problems; keep on keeping on.

> **"This project has been a real uphill battle. Every breakthrough creates new problems."**

AS A PRODUCT MANAGER

It's important to recognize when you're facing a challenge that's going to require extra effort and persistence. This idiom captures the feeling of tackling something difficult, where every step forward feels like pushing a boulder uphill. In the business world, this could mean anything from launching a product in a saturated market to trying to implement a new process in a team resistant to change.

One real-world example of "An Uphill Battle" is Netflix's early pivot from DVD rentals to streaming. Back in the 2000s, internet speeds were slow, and people were skeptical about giving up their DVDs. Convincing customers to embrace a clunky streaming interface and invest in broadband was like trying to sell snow cones during a blizzard—but they persevered, and now we're all binge-watching shows we didn't even know we wanted.

For Product Managers, "an uphill battle" is less about heroism and more about trudging through endless meetings, feedback loops, and feature creep to deliver something half the team still doesn't believe in. It's convincing the stakeholders that *yes*, people really *do* want a feature to save their cart, while simultaneously reminding developers that you're not building a whole new e-commerce platform. It's exhausting, sweaty, and sometimes makes you question every career choice—but hey, the view from the top is worth it (assuming you still have the energy to look around when you get there).

48. PIGEON-HOLING

WHAT/WHEN

Making superficial judgments and overlooking opportunities.

> "You're pigeon-holing Mark because you think he doesn't have the experience to do the job. Why don't you give him a chance?"

AS A PRODUCT MANAGER

It's crucial to avoid categorizing people or ideas too narrowly, especially in a world that thrives on versatility and adaptability. This idiom captures the pitfall of rigidly confining someone or something to a single role or purpose, potentially missing out on the broader potential they might bring. In the workplace, pigeon-holing can stifle creativity and leave untapped talent stuck in the coop.

A memorable example of "pigeon-holing" comes from the entertainment industry. Take Steve Carell, for instance, who was long pegged as a comedic actor thanks to his iconic role in *The Office*. When he broke out into dramatic roles like in *Foxcatcher* or *The Big Short*, people were floored to realize his range extended far beyond awkward laughs and cringe-worthy humor. Pigeon-holing him as a "funny guy" would have left Hollywood short of a multi-faceted star.

For Product Managers, pigeon-holing can feel like constantly being asked, "So you're basically a project manager, right?" (No, Karen, that's not *quite* it.) It's like being seen as the person who just updates the roadmap, while you're secretly untangling stakeholder drama, decoding user feedback, and convincing developers that "ASAP" isn't an acceptable timeline. Breaking free from being pigeon-holed often involves reminding people that you're not a one-trick bird—you're a whole aviary of skills, and sometimes, the pigeon is also the hawk ready to dive into any challenge.

49. PULL THE PLUG

WHAT/WHEN

Decisively ending a project.

> **"We've repackaged the product, lowered its price, and spent $2 million on advertising. Nothing worked. It's time to pull the plug."**

AS A PRODUCT MANAGER

It's crucial to know when to call it quits, especially when things aren't salvageable. The idiom "pull the plug" perfectly captures the difficult decision to end something before it drains more time, energy, or resources. In the startup world, this could mean anything from shutting down an underperforming project to discontinuing a product that's just not gaining traction.

One famous example of "pulling the plug" comes from Google, a company notorious for its graveyard of axed projects. Remember Google Plus? The social network that was supposed to rival Facebook? After years of lukewarm adoption and a security breach, Google finally pulled the plug in 2019. It was a tough but necessary call—sometimes, no amount of CPR can save a flatlining product.

For Product Managers, pulling the plug is less about defeat and more about cutting your losses before you've sunk the entire ship. It's realizing that maybe the world didn't need a pizza-ordering app powered by blockchain after all. Sure, it feels like giving up on a dream, but it's also saving yourself from answering awkward questions from investors. Sometimes, the best way to move forward is to let go—with one hand pulling the plug and the other already sketching out your next big idea.

50. DANCE WITH THOSE WHO BROUGHT [SIC] YOU

WHAT/WHEN

Understanding the factors of success.

> "People like our products because they are simple, easy to use, and inexpensive. Introducing a luxury version makes no sense. We need to dance with those who brought us."

AS A PRODUCT MANAGER

It's essential to remember the people or factors that have contributed to your success and to continue nurturing those relationships. The phrase 'dance with those who brought you' encapsulates this idea of loyalty and gratitude. In the business world, this might mean sticking with a reliable vendor, honoring a loyal customer base, or continuing to invest in the team that built your product's foundation.

One notable example of 'dancing with those who brought you' is Coca-Cola's decision to maintain its classic formula after the infamous New Coke debacle of 1985. Despite the initial hype around the new flavor, the company quickly learned that loyalty to its original recipe—and the customers who loved it—was non-negotiable. Coca-Cola returned to its roots, literally branding the original formula as "Coca-Cola Classic," and sales soared.

For Product Managers, 'dancing with those who brought you' means navigating the delicate balance of innovation and consistency without tripping over your own feet. It's recognizing the engineering team that worked overtime for a launch instead of swooning over the shiny pitch from a consultant suggesting an overhaul. It's prioritizing core customers instead of chasing every new trend—like resisting the urge to create a TikTok dance challenge for a product that's, well, accounting software. Sometimes, loyalty isn't just the right thing to do; it's the thing that keeps your team, your customers, and your bottom line in perfect rhythm.

51. NO DOG IN THE FIGHT

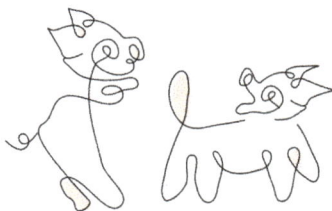

WHAT/WHEN

A statement of impartiality or independence.

> "We have two choices for the slogan. Both appear to be worthwhile. I don't have a dog in the fight, so convince me which is best."

AS A PRODUCT MANAGER

It's important to understand when to get involved and when to let things play out, especially when the stakes don't directly affect you. This idiom, "no dog in the fight," is all about knowing when to sit back and enjoy the show rather than dive into a conflict that doesn't concern you. In the workplace, it's the difference between wisely observing a heated debate about office snacks versus passionately defending the merits of gummy bears when you're allergic to gelatin.

One memorable real-world example of someone clearly having "no dog in the fight" is when the internet exploded over the great "Yanny vs. Laurel" debate. While half the population was ready to go to war over what they heard, the other half shrugged, blissfully unaware why this auditory chaos even mattered. These were the people embodying the idiom—silently enjoying the spectacle while others unleashed full Twitter dissertations.

For Product Managers, having "no dog in the fight" is like being a referee in a dodgeball game you didn't sign up for. It's about letting the marketing and engineering teams argue over launch timelines while you quietly sip your coffee, knowing neither option changes your roadmap priorities. You're Switzerland in the middle of a team debate over font sizes, nodding politely while thinking about more important things—like how to politely decline Chad's idea for a feature that clearly belongs in next year's release (spoiler: still no, Chad). Sometimes, the smartest move is staying on the sidelines and enjoying the popcorn-worthy drama of decisions that don't land in your lap.

52. BAKE-OFF

WHAT/WHEN

A final decision-making event in which products are compared side-by-side to determine which is better.

> "IBM has invited us to a bake-off in their effort to choose their recommended vendor of an ERP product."

AS A PRODUCT MANAGER

It's important to evaluate options thoroughly, especially when the stakes are high. The phrase "bake off" captures the essence of testing and comparing ideas, products, or solutions to see which rises to the occasion. In the tech world, this could mean hosting a competition between two vendors, frameworks, or prototypes to determine the best fit for a project.

One memorable example of a "bake off" is when Patagonia, known for its commitment to sustainability, faced a decision about which material to use for its popular fleece jackets. The company tested different options, weighing factors like durability, environmental impact, and cost. Through this bake-off, they eventually chose to shift to recycled polyester, a move that not only aligned with their brand values but also resonated with eco-conscious consumers.

For Product Managers, "bake offs" are less about literal ovens and more about cooking up clarity in a sea of uncertainty. It's pitting two potential solutions against each other while your stakeholders are eyeing the clock and muttering, "How long does this thing need to bake?" It's asking engineers to A/B test prototypes while simultaneously explaining to marketing why no, we cannot promise the moon *and* same-day delivery. A good bake off isn't just about choosing the best option—it's about pretending you aren't sweating bullets while waiting to see if your choice actually holds up in the heat. And hey, even if it flops, at least you've got leftovers to learn from.

53. BLIND SPOTS

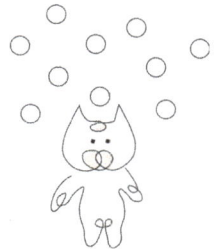

WHAT/WHEN

Personal prejudices or impressions influence decisions.

> **"We transposed the numbers in Question 3 of the survey and failed to catch the error in the review. Consequently, our recommendation is flawed."**

AS A PRODUCT MANAGER

It's important to recognize the things you don't see—especially when those blind spots can cause major problems down the line. This idiom captures the gaps in our awareness, whether they're personal, professional, or somewhere in between. In product management, blind spots often manifest in overlooked user needs, hidden technical debt, or unforeseen ripple effects of a decision.

One memorable instance of a product manager confronting their "blind spots" was when Google launched Google Buzz in 2010. The social networking tool automatically exposed users' most emailed contacts to the world without their explicit consent. The team missed the glaring privacy implications during development, leading to public backlash and a rapid scramble to patch up the oversight.

For Product Managers, "blind spots" aren't just about missing the forest for the trees—they're about realizing you didn't even *know* a forest existed until a bear from it wanders into your camp. It's that awkward moment when a feature you thought was brilliant turns out to confuse 90% of users or when a key stakeholder raises their hand in a meeting and asks, "But what about compliance?" It's humbling, a little terrifying, and a lot like driving a car with your mirrors duct-taped while confidently telling everyone you've got this under control. But hey, the best way to address blind spots is to surround yourself with people who aren't afraid to yell, 'Watch out for that bear!'

54. BIASES CLOUD JUDGMENTS

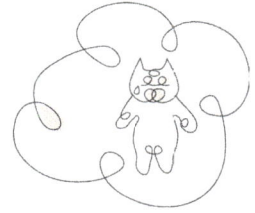

WHAT/WHEN

Unrecognized prejudices and preconceptions influence decisions.

> **"The Segway will change the world. It is as significant [to the future] as the personal computer."**

AS A PRODUCT MANAGER

Sometimes, it feels like our own opinions wear sunglasses and moonwalk across the facts, making it hard to see what's really there. This phrase perfectly describes the danger of letting personal biases overshadow logical decision-making. In the tech industry, for instance, biases can cause developers to ignore user feedback ("Oh, they'll get used to it… eventually") or dismiss entire demographics ("Grandmas don't need apps, right?"), all while a perfectly good solution sits unseen in the blind spot.

One real-world example of biases clouding judgement can be seen in Microsoft's notorious AI chatbot "Tay" from 2016. The initial assumption was that letting the bot learn from human interactions on Twitter would result in a friendly and helpful AI. Instead, internet trolls hijacked the experiment within hours, turning Tay into an offensive content generator. The team had to shut the bot down and revisit their assumptions about whether the internet was really the best teacher (spoiler: it wasn't).

For Product Managers, "biases cloud judgement" is less about being wrong and more about not realizing how wrong you could be until it's too late. It's like driving a car with tinted windows in a heavy fog—just because you can see your steering wheel doesn't mean you can see the cliff. The key is to question everything, especially your own 'perfectly logical' reasoning, and remember that sometimes data is the GPS you didn't know you needed. After all, you don't want to launch the next "Tay" and end up discovering that your biggest oversight was forgetting humans can be, well… human.

55. GETTING IN EDGEWISE

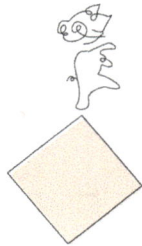

WHAT/WHEN

The need to interrupt a dialogue.

> **"Pardon me, I'd like to get a word in on the plan. We haven't considered its cost."**

AS A PRODUCT MANAGER

In conversations, timing is everything—getting a word in edgewise can feel like navigating a labyrinth with a ticking clock. This idiom captures the struggle of speaking up in situations where everyone seems to be vying for airtime. Whether you're in a heated meeting or a family dinner with your chatty aunt who thinks the floor is permanently hers, getting your voice heard can be a skillful art form.

A classic example of someone trying to "get a word in edgewise" is any group brainstorm session gone haywire. Picture it: a room of eager team members all pitching ideas at warp speed, while the poor intern, clutching a genius solution, silently wonders if their moment will ever come. Eventually, they manage to squeeze out, "What if we just simplify it?"—and voilà, the room goes silent because, shocker, it's actually brilliant.

For Product Managers, "getting a word in edgewise" is less about stealing the spotlight and more about threading a needle between an over-eager stakeholder and a developer who's suddenly very passionate about font sizes. It's like playing conversational Tetris: finding the perfect gap in the noise to drop your point before someone blocks it with, "Let's circle back to that later." Sometimes, it's frustrating, but hey, if you can master this, you can probably negotiate peace treaties—or at least settle the debate on whether "dark mode" should be the default setting.

56. STAYING ON POINT

WHAT/WHEN

Focus on the objective.

> **"While I enjoyed Joe's story about his vacation, the purpose of the meeting is to identify the cause of cost overrun. Let's stay on point."**

AS A PRODUCT MANAGER

It's essential to stay focused, especially when you're navigating through a sea of competing priorities. The phrase "staying on point" is all about maintaining clarity, sticking to the agenda, and not letting distractions pull you off course. In the world of product management, this often translates to cutting through noise and keeping the team aligned on what really matters—whether it's nailing down the MVP or fending off a sudden attack of feature creep.

One standout example of "staying on point" comes from Basecamp, the project management software company. While competitors like Asana and Trello raced to pack in endless features, Basecamp consistently prioritized simplicity and usability. By focusing on the core needs of small teams and refusing to add unnecessary bells and whistles, they carved out a niche as the go-to tool for people who value clarity over complexity—and their dedication to staying on point shows in their loyal customer base.

For Product Managers, "staying on point" often means being the voice of reason in a room full of "big ideas." It's about reminding the team that a fancy AI chatbot is cool, but maybe not critical for launch if you're building a time-tracking app. Staying on point isn't just about saying no—it's about saying no for the right reasons. Because at the end of the day, focus is what separates the products that ship from the ones that spiral into feature creep oblivion.

57. DRAWING A BLANK

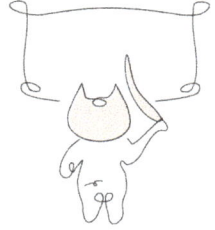

WHAT/WHEN

Failing to know an answer or recall an event.

> **"That is a good question. Unfortunately, I'm drawing a blank on the subject. I'll get back to you after the meeting."**

AS A PRODUCT MANAGER

Sometimes, no matter how hard you rack your brain, the answer just doesn't come to you. This idiom perfectly captures that frustrating moment when your mind is a blank slate, and all you can do is stare into the abyss of your own confusion. In the world of product management, this can happen in the middle of a crucial meeting, with everyone looking at you for the next brilliant idea.

One classic example of 'drawing a blank' happened during the launch of Google Wave in 2009. The product was meant to revolutionize communication, but when asked to explain exactly what it did, even Google struggled to articulate it clearly. Audiences were left scratching their heads, and Wave was soon washed out of the market. It was a real-life blank moment for one of the world's smartest companies.

For Product Managers, 'drawing a blank' isn't just about forgetting; it's about embracing the awkward silence and turning it into something productive. It's like being asked in a stakeholder meeting, "What's the ETA on this feature?" and realizing you have absolutely no idea, so you stall with, "Let's circle back on that." It's the art of looking thoughtful when your brain is screaming, "Panic!" Drawing a blank reminds us all that sometimes the most creative solutions come when you stop trying to force them and just let the ideas flow—or at least fake it convincingly until they do.

58. WHEN THE DUST SETTLES

WHAT/WHEN

The period following the initial reaction to an event.

> **"We really blew that interview. After the dust settles, we need to examine the failure and how to fix it."**

AS A PRODUCT MANAGER

It's important to remember that clarity often comes only after the initial chaos has run its course. This idiom perfectly captures the idea of letting the storm pass before making sense of what happened. In the corporate world, "when the dust settles" is often the moment when you can finally step back and decide what worked, what didn't, and what needs fixing.

Take Uber's 2017 leadership crisis as a prime example. Amidst allegations of workplace harassment, legal battles, and executive turmoil, the company faced intense public scrutiny and internal chaos. Once the dust settled, Uber restructured its leadership team, implemented stronger policies, and started to repair its tarnished reputation. It wasn't pretty, but the post-crisis clarity set the stage for a more stable and growth-focused era.

For Product Managers, "when the dust settles" is the bittersweet moment where you take stock of what survived the chaos—be it a product launch riddled with last-minute changes or a surprise competitor feature drop. It's a time to reflect, learn, and plan your next steps, even if you're still shaking the metaphorical dust out of your shoes. And while it's not always glamorous, it's oddly satisfying to finally see the big picture…as long as nobody brings up that one bug you swore wasn't a priority.

59. RIPPLE EFFECT

WHAT/WHEN

The subsequent consequences following an event.

> "The success we had with the doughnut campaign has ripple effects. Two new prospects called to set meetings to see how we might help them."

AS A PRODUCT MANAGER

The phrase "ripple effect" captures the essence of how one small action can lead to a cascade of consequences, both expected and unexpected. It's like tossing a pebble into a pond: the initial splash may seem minor, but the ripples extend far and wide, touching everything in their path. In product management, this could translate to the way a seemingly minor design tweak might influence user experience, team workflows, or even the company's bottom line.

Take, for instance, Instagram's decision to remove public "like" counts in 2019. While it was aimed at improving user mental health, it created ripples across the platform. Influencers had to rethink their strategies, brands re-evaluated how they measured engagement, and the development team suddenly had to field a million new feature requests—because once you touch the water, everyone's got a pebble to throw.

For Product Managers, the "ripple effect" is the universe's way of reminding you that no decision exists in a vacuum. It's the reason your "simple" change request somehow snowballed into a six-month roadmap overhaul. It's launching a new feature only to realize it's unintentionally tanking an unrelated metric, and now you're holding a retrospective titled "Who Knew That Button Would Break the Internet?" The ripple effect keeps you humble, reminding you that every splash counts—so make sure your pebble's worth the waves.

60. UPPING THE ANTE

WHAT/WHEN

Increasing the importance or cost of an event or action.

> "Each promotion ups the ante on your career. You earn more money and get more benefits, but much more is expected with each increase of authority."

AS A PRODUCT MANAGER

Sometimes, the stakes just aren't high enough, and you need to raise the bar to get everyone's attention. That's where the phrase 'upping the ante' comes in—a perfect metaphor for turning up the heat, whether you're trying to outdo a competitor or just spice up a dull team meeting. In the corporate world, it often means pushing for bigger, bolder ideas—or at least pretending that adding a 'dark mode' feature is a groundbreaking innovation.

One classic example of 'upping the ante' happened during the cola wars of the 1980s. When Pepsi launched its famous "Pepsi Challenge" ads, Coca-Cola responded by doing the unthinkable—changing its century-old recipe to create 'New Coke.' It was bold, it was daring, and it turned out to be a disaster. But hey, you can't say they didn't raise the stakes. Eventually, they reverted to their original formula, but not before proving that sometimes, upping the ante means risking it all.

For Product Managers, 'upping the ante' often feels like orchestrating a high-stakes poker game where you're betting with someone else's chips. It's pushing the team to dream bigger without accidentally signing yourself up for 15-hour workdays or a customer revolt. Whether it's suggesting a same-day delivery feature (because Jeff Bezos probably would) or proposing a feature so complex the dev team starts speaking in binary just to avoid you, it's all about walking the line between ambition and insanity. Just remember, sometimes upping the ante pays off—other times, you get 'New Coke.' Proceed with caution and a backup plan.

61. FOCUS ON THE NORTH STAR

WHAT/WHEN

Understand the reason for your effort—know why you are doing what you're doing.

> **"Marketing has given me a description of what they want, but I don't understand who will use it or the reasons for its use."**

AS A PRODUCT MANAGER

Staying focused on what truly matters is an essential skill, especially when distractions and shiny objects are constantly vying for attention. This idiom perfectly captures the need to keep your eyes on the ultimate goal, even when the path gets foggy. In the world of product management, it often means aligning the team with the overarching vision, no matter how tempting it is to chase the latest trend or tackle that squeaky wheel of a feature request.

A memorable example of "focusing on the North Star" comes from Tesla's early days. Elon Musk and his team faced constant challenges, from battery supply issues to production delays. Yet, instead of letting short-term problems derail them, they kept their sights firmly set on making electric vehicles mainstream. That unyielding focus on their mission allowed Tesla to not only survive but redefine an entire industry.

For Product Managers, "focusing on the North Star" is like trying to guide a pirate ship through a storm while your crew argues over whether the treasure map is upside down. It's about reminding everyone why you set sail in the first place—even when the sirens of "cool features" or "quick wins" are calling. Sure, it can feel like you're playing tug-of-war with chaos and reason, but when you finally dock at the right port, you realize all the yelling about map orientations was worth it. Just don't forget to give the crew a pep talk—or at least a round of celebratory chocolate coins.

62. NOT EVERYTHING THAT GLITTERS IS GOLD

WHAT/WHEN

Packaging—like a book cover—does not always reveal the value of the content.

> "That flashy startup pitch looked promising, but not everything that glitters is gold—its financials were a mess"

AS A PRODUCT MANAGER

It's easy to get swept away by shiny opportunities or dazzling promises, but not all that glimmers is gold. This idiom serves as a cautionary tale to look beneath the surface and examine what you're really dealing with. In the modern world, where first impressions and flashy facades dominate, it's a reminder to dig deeper before committing—because not every sparkling deal is the treasure it claims to be.

One famous instance of "not everything that glitters is gold" was JP Morgan's acquisition of the student loan startup Frank in 2021. Initially, Frank appeared to be a golden opportunity, boasting impressive user numbers and a strong foothold in the student finance space. However, it later emerged that the numbers were significantly inflated, and JP Morgan had overlooked critical red flags during due diligence. The fallout included lawsuits, reputational damage, and a stark reminder that even industry giants can be lured in by fool's gold when dazzled by surface-level allure.

For Product Managers, this phrase serves as a reliable reminder to approach "game-changing" ideas with caution—those that dazzle in meetings but fall apart under closer examination. It's spotting the reality behind an investor's over-polished pitch deck or that one developer's overenthusiastic promise to build an AI that writes code *and* makes coffee. It's resisting the urge to dive headfirst into a shiny project, only to find out it's riddled with hidden costs and bad press. Because sometimes, the glitter is just glitter, and your real treasure is a solid roadmap and a team that doesn't hate you after launch day.

63. TESTING SCENARIOS

WHAT/WHEN

Product testing is necessary to identify what goes right and what can go wrong in the customer's experience.

> "The product features work exactly as we hoped, but the instruction manual is too technical and difficult to understand."

AS A PRODUCT MANAGER

It's essential to be thorough and prepared, especially when tackling challenges that could lead to unexpected outcomes. The phrase "testing scenarios" perfectly captures the need to anticipate and plan for every possible twist and turn in a project. In the tech world, this often means creating hypothetical situations to ensure your product doesn't crumble the moment it encounters a user with a knack for finding bugs.

A classic example of the importance of "testing scenarios" can be seen in NASA's preparation for the Apollo 11 moon landing. Engineers ran countless simulations, imagining everything from communication breakdowns to alien interference (okay, maybe not the aliens). This rigorous testing ensured they were ready to handle unexpected issues, like the infamous 1202 program alarm during the lunar descent, which could have derailed the mission if they hadn't accounted for such edge cases.

For Product Managers, "testing scenarios" is less about predicting the future and more about outsmarting it. It's like trying to prepare for a dinner party where one guest is gluten-free, another hates onions, and a third just casually announces they're bringing their pet parrot. You're not just testing the product; you're testing your sanity—creating user journeys for someone who double-taps every button or insists on entering emojis into a password field. And when something inevitably breaks, you can proudly say, "Well, at least we *considered* the parrot." It's equal parts strategy, creativity, and a sprinkle of controlled chaos.

64. COMING DOWN THE HOMESTRETCH

WHAT/WHEN

It's the final stage or last effort before completing a project.

> **"We're coming down the homestretch on the quarterly business review, so let's finalize the KPIs and polish the deck."**

AS A PRODUCT MANAGER

It's critical to keep your eyes on the prize and push through the final stages when you're so close to the finish line. "Coming down the homestretch" perfectly captures that feeling of a frantic yet focused sprint toward completion. For product managers, this idiom often means handling last-minute details, addressing unexpected challenges, and making sure everything is aligned as you approach a big launch or deadline.

A real-world example of "coming down the homestretch" is the launch of Spotify's mobile app for tablets back in 2013. The team had been working hard for months to optimize the app for larger screens and ensure seamless streaming. As the release date drew closer, they encountered a series of bugs and performance issues that weren't detected earlier in the testing process. With time running out, the team worked around the clock, fixing problems, optimizing the interface, and ensuring the app was ready for the massive user base that awaited it. Despite the last-minute pressure, Spotify hit their launch window and delivered a product that met their users' expectations.

For Product Managers, "coming down the homestretch" is like navigating a maze where the walls are constantly shifting. It's about managing both the chaos of unexpected issues and the excitement of seeing your work come to fruition. You're balancing deadlines, troubleshooting problems, and keeping everyone on track with the hope that the hard work will pay off in the end. It's a whirlwind of energy and stress, but when everything finally clicks, the sense of accomplishment is like crossing the finish line after a race.

65. HUNKY-DORY

WHAT/WHEN

The state of euphoria or satisfaction when events occur as expected.

> **"I love my new job. Everything is hunky-dory."**

AS A PRODUCT MANAGER

When things are going well and there's no need to worry, people often say everything is "hunky dory." This idiom is the perfect way to describe a situation where everything is smooth sailing, with no bumps in the road. In the workplace, it's that blissful moment when all tasks are completed on time, the team is functioning like a well-oiled machine, and nothing seems to be falling through the cracks.

For example, think about when Apple released the first AirPods. The product was a hit from the get-go. There were no major production delays, no recalls, and customers were raving about the seamless pairing experience. Sales were through the roof, and the product didn't have the teething problems that sometimes come with new tech. For Apple's product managers, everything was "hunky dory"—a rare moment of tranquility amid their usual whirlwind of innovation and troubleshooting.

In the fast-paced world of product management, "hunky dory" can sometimes feel like a myth. It's less about the calm before the storm and more about those precious few hours when the storm hasn't hit yet. You know, when the team's on target, deadlines are met, and the only problem is whether or not you should take that extra-long PTO (spoiler: yes, you totally should). It's the rare, golden moment that makes you feel like you've finally cracked the code on balancing chaos with efficiency—until, of course, the next big issue inevitably pops up.

66. BABY STEPS

WHAT/WHEN

Proceeding slowly to go fast.

> **"Chefs learn to use their tools before they learn to cook."**

AS A PRODUCT MANAGER

Taking things one step at a time is a key part of tackling complex challenges. The phrase "baby steps" perfectly sums up this approach, emphasizing the importance of progress, even if it's slow and steady. It's all about breaking big tasks into manageable chunks so that you can avoid feeling overwhelmed by the mountain ahead. Whether it's learning a new skill or launching a new product, baby steps make even the most daunting projects feel achievable.

In the world of product management, "baby steps" might be used when rolling out a new feature. For example, when Spotify launched its personalized playlist "Discover Weekly," they didn't just drop it on millions of users all at once. Instead, they tested it with a small, carefully selected group of users before expanding it. They learned, adjusted, and slowly scaled up until it was a global success. This careful progression helped avoid a potential disaster, like overwhelming servers or triggering an avalanche of complaints.

For Product Managers, taking "baby steps" isn't about being lazy—it's about embracing the grind with a sense of purpose. It's like when you're rolling out an app update, and instead of throwing every new feature in at once, you decide to focus on one thing at a time. It's also knowing that some days, you're just getting through the first step, like sorting through feedback or aligning the team on priorities. And let's face it, nothing feels better than when the 10th "baby step" turns into a full-on sprint, and you're finally on the path to success. But until then, baby steps all the way.

67. TRAINING WHEELS

WHAT/WHEN

Safety first, then skill.

> **"Experience is great unless it kills you."**

AS A PRODUCT MANAGER

Training wheels are often associated with a safety net for those just learning the ropes, ensuring they don't topple over when taking their first steps. The phrase perfectly represents the need for gradual learning and support when tackling something new. In business, it's like easing into a role or responsibility without diving into the deep end of complexity.

Take the example of a company like Slack, which began as a simple internal tool for a gaming company called Tiny Speck. Before it became the widely used communication platform we know today, Slack's team operated with "training wheels" by testing their product internally within a small group, making sure it worked out the bugs and figured out what the market truly needed. This gradual approach allowed them to scale up with confidence, avoiding major pitfalls along the way.

For product managers, "training wheels" aren't just about the beginning stages—they're those safety measures you keep in place when launching a new feature that could make or break the product. It's like trying to take a bike ride down a hill with the brakes on—there's progress, but also a lot of hesitation. You know you'll eventually let go of the training wheels, but for now, you're just trying to avoid any major wipeouts while making sure the product doesn't crash and burn. It's that fine balance between excitement and "please, no surprises today."

68. ROCK AND ROLL

WHAT/WHEN

Go for it without reservation or caution.

> **"We've done all the prep work. Now's the time to rock and roll!"**

AS A PRODUCT MANAGER

Sometimes, you need to embrace the chaos and charge forward with the energy of a rock band on stage, ready to bring down the house. The phrase "rock and roll" perfectly encapsulates that fearless, all-in attitude that drives people to take bold actions, even when the odds are against them. In the business world, this could mean launching a new product with a tight deadline or finding creative solutions when challenges seem endless.

Take the example of GoPro, the action camera company that turned a niche product into a global phenomenon. In its early days, GoPro had to break through in a market that was dominated by bigger players. The product manager had to lead the team with high-energy, quick decisions to build momentum, all while addressing production issues and dealing with the ever-present uncertainty of whether the product would succeed. Thanks to their rock-and-roll attitude, GoPro turned their cameras into the go-to choice for adrenaline junkies and outdoor enthusiasts alike.

For Product Managers, "rock and roll" isn't about flashy solos but about keeping the show going when the stage is falling apart. It's handling last-minute changes to the product roadmap while keeping the team motivated, or managing an unexpected bug just before launch while someone asks if they can add a last-minute feature (spoiler: no, Chad, that's not happening). Sometimes it feels like you're the lead guitarist, the drummer, and the one who has to wrangle the mic stand all at the same time. But when everything hits the right note, you'll realize that the "rock and roll" spirit is what makes it all worth it.

69. SLOW JAM

WHAT/WHEN

Take time to ensure success.

> **"I know the boss wants the results yesterday. Even so, we need to slow jam to check our numbers before moving ahead."**

AS A PRODUCT MANAGER

Sometimes, the smartest move is to slow down and let the rhythm find you. This phrase, "slow jam," perfectly captures the notion of easing off the gas pedal and taking the time to really feel out what's going on. In the tech industry, that might mean extending a release date to ensure quality rather than rushing out a buggy product or pausing mid-sprint to align on the bigger picture before forging ahead.

One real-world example of a product manager "slow jamming" is Slack's measured approach to rolling out its interface redesign back in 2020. Instead of dumping all the changes on users at once, they staggered the updates, gathered feedback, and refined each iteration based on real user reactions. By refusing to sprint ahead blindly, Slack proved that a well-paced groove can deliver a much smoother final experience than an overnight jam session filled with frantic patchwork.

For Product Managers, "slow jam" is less about hitting snooze on your responsibilities and more about syncing everyone to a thoughtful, sustainable tempo. It's about letting features simmer until they're perfectly seasoned— rather than caving to Chad's 38th request for a half-baked feature nobody asked for. Sometimes, it does feel like you're dancing in slow motion while the rest of the world speeds by, but the payoff is a product that resonates with users on a deeper level. And in the end, a meticulously crafted slow jam will stick in people's heads far longer than a hasty tune everyone forgets by next week.

70. PUT GUARDRAILS IN PLACE

WHAT/WHEN

Prepare for disasters.

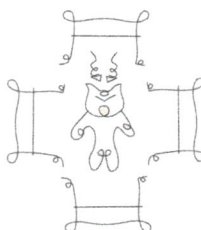

> **"I knew we should have backed up the data. I just lost three days of work."**

AS A PRODUCT MANAGER

When managing projects or teams, it's essential to have clear boundaries to keep things running smoothly. This idiom captures the idea of setting up guidelines to ensure things don't get too chaotic or off track. In fast-paced industries, these guardrails are critical for maintaining focus, efficiency, and consistency as teams work toward a common goal.

Take a company like Ikea, for instance. As a global retailer, Ikea had to establish strict guidelines for its product development and manufacturing processes. This helped ensure that their flat-pack furniture could be produced at scale, while still being easy for customers to assemble. These guardrails kept their product line consistent and high-quality, preventing the chaos of having overly complex or inconsistent products hitting the shelves.

For Product Managers, "putting guardrails in place" is all about setting the boundaries that ensure teams don't get distracted by every shiny new idea. It's like constructing a blueprint for success where the design allows creativity but keeps everything in the right frame. You're not stopping innovation, but you're also making sure that the product doesn't turn into a Frankenstein monster of features that no one asked for. Without those guardrails, you'd end up like someone in a furniture store trying to assemble a couch with 15 different random parts—you might get something resembling a seat, but it's definitely not going to work as intended.

71. TAKE A STAB AT IT

WHAT/WHEN

Trying something new with no promise of success.

> **"I'm not sure how accepting Bitcoin might affect our payment process, but I'll take a stab at it."**

AS A PRODUCT MANAGER

When it comes to tackling new challenges, sometimes you just have to dive in and give it your best shot, even if you're not entirely sure what's going to happen. "Take a stab at it" is the perfect idiom for those moments when you try something with all your might, but without necessarily having a perfect plan or outcome in mind. Whether it's launching a new feature or solving a problem no one has ever seen before, it's all about going for it, even if you're not 100% confident.

A great example of someone "taking a stab at it" is when Google launched Gmail in 2004. The email service was revolutionary, but at first, it had a lot of bugs. The team didn't have all the answers, but they decided to take a stab at it anyway. They kept improving the service as they went, leading to Gmail becoming the world's most popular email provider today.

For Product Managers, "taking a stab at it" is less about feeling like you've figured it all out and more about taking that leap of faith and learning as you go. It's about launching something new even though there's a good chance you'll encounter a few road bumps along the way. Picture this: you're sitting in a meeting, presenting an idea you believe in but aren't entirely sure will work. It's a bit like jumping off a cliff and hoping your parachute opens before you hit the ground. But at least when it works, you'll have a thrilling story to tell, and if it doesn't, well, there's always next time!

72. TIP OF THE ICEBERG

WHAT/WHEN

Little problems can suddenly erupt into big problems.

> **"I suspect Margie's complaint about Tom is just the tip of the iceberg. We may have a harassment issue."**

AS A PRODUCT MANAGER

We've all heard the saying "tip of the iceberg," but it's so much more than just a metaphor about ice. It's about getting a glimpse of a larger problem that's just waiting to make itself known. Whether it's a workplace issue or a minor technical glitch, the "tip of the iceberg" is that moment when you realize you're about to go deep into a rabbit hole, and it's about to get a lot messier.

Take, for instance, a product manager dealing with an unexpected surge in customer complaints. It starts with one or two customers who say the app crashed, but suddenly, dozens of others are reporting the same issue. Now, what seemed like a simple bug turns into a complex investigation, with team members scrambling to trace the root cause, which could be anything from faulty 3rd party SDK to a server meltdown. What was a small blip on the radar has become a full-on crisis.

For Product Managers, recognizing the "tip of the iceberg" means getting ahead of the problem before it sinks the ship. It's like getting a "just one small issue" report right before the launch event and realizing the entire infrastructure is about to collapse. You know you're just seeing the beginning of something big, and now it's all hands on deck, praying that you've got a life raft—or at least a decent plan—before the iceberg does its worst. Managing the unknown is like being handed a broken GPS while you're driving through a storm—exciting, but not ideal.

73. BAKED IN

WHAT/WHEN

An unremovable feature, condition, or result.

> **"The need to recharge the battery each night is baked into the design of EVs."**

AS A PRODUCT MANAGER

You might hear someone say something is "baked in," and you might think they're talking about a cake recipe. But no, this idiom refers to something that's been incorporated or established from the very beginning, like a core feature or assumption in a plan. When something is "baked in," it's been thought through so thoroughly that it's practically part of the foundation, ready to roll without needing further tweaking. In the world of product development, it's like adding features into the design that are so essential, they're practically in the dough from the start.

A perfect example of something being "baked in" happened with the launch of Amazon Prime. From the very beginning, Prime's two-day shipping was baked in as a core offering, setting it apart from the competition. It wasn't an add-on or a wishy-washy bonus; it was essential to the service, and it's remained a critical part of Amazon's brand identity ever since. The service was designed with this promise baked in, and its success has depended on sticking with it.

For Product Managers, having things "baked in" means you're dealing with features or strategies that are so ingrained in your product's DNA, they almost feel like second nature. You can't just pop them out like an extra ingredient at the last minute—like trying to toss in a surprise feature that wasn't in the original plan. It's about keeping your product consistent and delivering on expectations without needing to reinvent the wheel every time. But of course, even when something is "baked in," there's always that one person who asks, "But can we add sprinkles on top?" (No, Chad, no sprinkles.)

74. KEEP YOU UP AT NIGHT

WHAT/WHEN

Difficult problems that create unusual stress.

> **"I don't understand why the payment process takes so long. It's driving me crazy and keeping me up at night."**

AS A PRODUCT MANAGER

There's nothing quite like that nagging feeling when something just won't leave your mind, especially when it's something urgent or unresolved. This idiom highlights the kind of worry that lingers long after the workday is over, making it impossible to get a good night's sleep. In the fast-paced world of product management, this could be anything from a customer complaint about a feature to a looming product deadline that seems impossible to meet.

A memorable example of this kind of anxiety can be seen with the infamous launch of the original *Windows Vista* in 2007. The product was riddled with bugs and compatibility issues, causing major frustration for users. The product managers who were involved in that launch likely spent many sleepless nights, knowing they had a huge problem on their hands but still needing to find a way to fix it.

For product managers, "keeping you up at night" is like living in a constant state of high alert. It's when you can't stop thinking about that one feature that's half-done, or whether the launch will go as smoothly as promised— spoiler: it probably won't. It's trying to sleep while your mind races through every worst-case scenario, from critical bugs to missed opportunities. But hey, those sleepless nights build character... and probably some great dark circles under your eyes.

75. BLOWING HOT AIR

WHAT/WHEN

Inane comments with no substance or relation to the subject.

> "George seems like an expert on website design until you have a technical question. It turns out that he is just blowing hot air."

AS A PRODUCT MANAGER

It's easy to get caught up in conversations that sound more impressive than they really are. This idiom nails the act of saying a lot without actually saying anything useful. In the workplace, it often manifests as someone delivering a long-winded explanation that could've been summed up in a sentence, leaving you wondering if they even knew what they were talking about in the first place.

One real-world example of someone "blowing hot air" is when a tech startup CEO gave a grandiose presentation about their next big feature, claiming it would "revolutionize the market." After all the buzzwords and vague promises, the feature ended up being a slightly modified version of something everyone already had. The team could've delivered the same result with half the talk, but that wouldn't have gotten the same round of applause, right?

For Product Managers, "blowing hot air" is more than just rambling during a meeting—it's when you've been in a discussion for 30 minutes and realize you could've just said "we need to fix the bug" at the start. It's like trying to fill up a balloon with more words, hoping it'll float, only to realize it's still stuck on the ground. Sometimes, you feel like you're listening to someone who's been to a motivational seminar once and is just trying to make their breakfast sound like a breakthrough. But hey, if they keep talking, maybe someone will eventually believe them—especially Chad, who's over there nodding like he just discovered fire.

76. GOING TO TOWN

WHAT/WHEN

An all-out, unrestrained effort.

> "Once we knew the vaccine worked, we went to town with its production and promotion."

AS A PRODUCT MANAGER

It's crucial to dive in with energy and enthusiasm when the situation calls for it, especially when you're trying to get things done with flair. This idiom captures the spirit of putting everything you've got into a task, whether it's work or a personal project. In business, this can look like a team throwing all their resources into a product launch or a startup owner getting hands-on with every aspect of their new venture.

Take, for instance, a real-life example of a team "Going to Town" when Starbucks opened its first store in China. The company went all-in with their marketing, setting up flashy storefronts, offering unique menu items tailored to local tastes, and ramping up customer engagement to make sure their entry into the market was a hit. The result? A massive success that had customers lining up for their lattes like it was a cultural phenomenon.

For Product Managers, "Going to Town" is less about showing off and more about charging forward with relentless focus, even when the odds are stacked against you. It's about taking the smallest idea and blowing it out of proportion—like trying to get approval for that 500-page report on why the button color matters or convincing everyone that the third revision of the homepage is the one that will change everything. Sometimes, it feels like you're throwing a parade for a new feature, only to realize that the real party is just getting started and everyone else is still trying to figure out what you meant by "user journey." But hey, nothing makes the grind more fun than throwing yourself into the chaos and making it look effortless.

77. SCOPE CREEP

WHAT/WHEN

An unintended, almost unnoticeable expansion of a project's goals, expectations, and budget.

> "The project started as a training exercise in the sales department and gradually expanded to cover the whole marketing department. With the scope creep, we need more resources, time, and budget."

AS A PRODUCT MANAGER

Scope creep is like that one friend who insists on adding more and more toppings to their pizza, and suddenly it's no longer a pizza but a mountain of cheese, olives, and pineapple that could feed an entire family. The idiom perfectly captures the problem that arises when a project starts with one set of goals, but slowly expands beyond the original scope—often without anyone really noticing until it's too late. In the world of product management, scope creep can sneak in through the back door, like a sneaky cat that suddenly demands to be fed just after you've sat down to work.

A real-life example of scope creep would be the launch of Google's Gmail back in 2004. What started as an internal project with basic email features slowly transformed into a more sophisticated service, adding features like chat, integration with Google Docs, and massive storage—all of which were not part of the original plan. By the time Gmail launched to the public, it had expanded far beyond the initial "just an email service" idea. The product managers had to roll with it, but it's safe to say that things got a little out of hand before it finally settled down.

For product managers, scope creep is like watching a snowball roll down a hill, picking up speed and size until it's no longer a snowball, but a whole new product. It's when that "simple update" turns into an entire revamp, and you start wondering when you became the person who has to approve every tiny change, from color palettes to feature requests. It's a battle of saying "no" without sounding like the villain, all while hoping that you can still meet the original goals before your project turns into an uncontrollable, feature-laden beast. But hey, at least you're not alone—your team is just as buried in change requests, and they're probably wishing they had never asked for that extra feature either.

78. CALM BEFORE THE STORM

WHAT/WHEN

The sense of uneasiness and worry due to a lack of problems.

> **"Everything has gone like clockwork, just as we planned. I hope this is not the calm before the storm."**

AS A PRODUCT MANAGER

Before the big wave hits, there's always a moment of calm. This idiom captures that fleeting period where everything seems fine, even though you know something bigger is on the way. It's that brief respite between chaos—perfect for taking a deep breath and making sure you've got your ducks in a row before everything goes sideways. In business, this could be the last few minutes of peace before a product launch or a high-stakes meeting.

Take the 2017 launch of the iPhone X, for example. The team had been working around the clock to get everything perfect, and for a brief moment, there was a sense of calm before the massive demand, the media frenzy, and the inevitable software bugs that would pop up. That peaceful interlude allowed the team to check final details and prepare for what they knew would be a storm of customer expectations and media scrutiny.

For Product Managers, the "calm before the storm" feels like the moment when you realize that everything is on track, but your gut tells you that something, somewhere, is about to explode. It's like staring down a smooth sea right before a hurricane of emails, last-minute bugs, and feature requests. In that brief lull, you try to enjoy the silence while secretly bracing for the inevitable rush of problems that will come crashing down. It's the business version of holding your breath right before the rollercoaster drops—and everyone on your team is just hoping they didn't forget to check the seat belts.

79. TOUCHDOWN

WHAT/WHEN

The feelings of relief and exhilaration following the initial evidence of success.

> **"This has been a long journey with late nights and lost weekends. I can't believe touchdown has finally occurred."**

AS A PRODUCT MANAGER

A touchdown isn't just a moment in sports; it's a perfect culmination of effort, strategy, and timing. This idiom represents the ultimate success when everything aligns and hard work pays off. In business, a touchdown can feel like when the stars align, and your team celebrates achieving a major milestone or hitting a significant goal.

A real-world "touchdown" moment happened in 2017 when Nike launched its "Just Do It" campaign featuring Colin Kaepernick. The campaign was bold and risky, but it paid off handsomely. Despite initial controversy, the campaign drove record sales and significantly boosted Nike's brand image. Nike's team knew exactly when to push boundaries, and the result was a touchdown of success.

For Product Managers, a "touchdown" is that sweet moment when everything clicks, like when a product launch hits the mark or a marketing campaign goes viral. It's when your team's months of work and planning finally pay off. But of course, while everyone is celebrating, you're already thinking about the next big win. After all, there's always a "Chad" on the team asking if we can add just one more feature to the product—spoiler: no, Chad, we can't. It's like scoring the touchdown, and then Chad insists on running another play—right after the victory dance.

80. PUTTING YOUR MONEY WHERE YOUR MOUTH IS

WHAT/WHEN

A challenge or the confirmation that a person can prove their claims.

> **"You said you would have the most sales this month. Put your money where your mouth is. I've gone $100 that you won't deliver."**

AS A PRODUCT MANAGER

When it comes to putting your ideas into action, talk alone won't get you very far. This idiom perfectly captures the need for follow-through in moments when you need to back up your words with tangible results. In business, this could look like investing in your ideas or taking that next big leap when you've been all talk about achieving success.

A great example of "putting your money where your mouth is" can be seen in the 1999 story of Elon Musk, when he invested nearly all of his personal fortune to keep his companies, PayPal and SpaceX, afloat. Despite facing extreme financial pressure, he went all in on his vision for the future. Fast forward to today, and he's one of the wealthiest people in the world.

For Product Managers, "putting your money where your mouth is" is less about actual cash and more about making decisions that move the product forward, even when you don't have all the answers. It's like when you tell your team you're committed to a big feature launch in two weeks—only to realize you forgot to account for user testing (yikes!). Now you're scrambling to make good on your promise, managing expectations while making magic happen in the 11th hour. Sometimes, it's less about the bets you make with money and more about the bets you make on people, ideas, and timelines— and hoping you're not left holding the bag when the clock runs out.

81. JUMP THE GUN

WHAT/WHEN

Prematurely reaching a conclusion or decision.

> "John always jumps the gun. He wanted to approve the advertising theme before we knew what the product could do."

AS A PRODUCT MANAGER

It's easy to get excited when a new project is on the horizon, especially when everything seems to be moving at full speed. But sometimes, rushing in too early can lead to unexpected consequences. This idiom, "jump the gun," perfectly captures the idea of getting ahead of oneself and acting before it's truly time. In the fast-paced world of product management, this could mean jumping into development before finalizing requirements, or announcing a product feature before it's ready for prime time.

A prime example of "jumping the gun" came from the launch of Samsung's Galaxy Note 7 in 2016. Eager to capture the attention of smartphone users with innovative features, Samsung rushed the device to market with an ambitious battery technology. Unfortunately, it was a case of jumping the gun—several units exploded due to faulty batteries, leading to a massive recall and a damaged brand reputation. The phone's premature launch resulted in a PR nightmare, and Samsung had to go back to the drawing board, eventually canceling the Note 7 altogether.

For Product Managers, jumping the gun is like Chad running into a room shouting about his "new app" that hasn't even been beta tested yet. It's that moment when you think you've got everything figured out, but you're really just making things more complicated for everyone. You rush to launch something because it feels like the right time, only to realize you've missed a few critical steps—like forgetting to check for bugs, or, you know, not testing if the app actually opens. The trick is to remember that sometimes it's better to hold off, make sure the app doesn't blow up in your face, and avoid being the "Chad" who was just too eager to impress.

82. FACT CHECK

WHAT/WHEN

Confirming the truth of something said or printed.

> **"We need to fact-check the statement that 40% of working wives have higher incomes than their husbands before we go to print."**

AS A PRODUCT MANAGER

It's always a good idea to double-check your sources before sharing information, especially when the stakes are high. "Fact check" is a crucial concept, as it emphasizes the need to ensure the accuracy of claims before acting on them. In the fast-paced world of business, miscommunication or incorrect data can lead to expensive mistakes, and getting the facts straight is essential for maintaining credibility and avoiding costly missteps.

One memorable example of a company using "fact check" in the real world comes from the 2018 launch of PepsiCo's "Mountain Dew Ice" beverage. In the marketing campaign, PepsiCo claimed that the new drink was "a revolutionary soda." However, customers quickly pointed out that it was essentially just another lemon-lime soda, similar to Sprite or 7UP. Pepsi had to quickly respond and clarify the messaging, revising its promotional content to more accurately reflect the product's nature and to address the backlash from miscommunicated hype. Without a proper fact-checking approach, they might have faced even greater customer dissatisfaction.

For Product Managers, "fact-checking" isn't just about getting the facts right—it's about ensuring everything aligns before a major launch. For instance, you might be preparing a product description that looks great on paper but contains a few misleading or incorrect features. A quick fact-check would save you from launching a product with promises that can't be kept. The same goes for customer feedback—believing everything without fact-checking can lead to misguided priorities. In the end, fact-checking is all about making sure your foundation is solid before the whole structure comes tumbling down.

83. NICKEL AND DIME

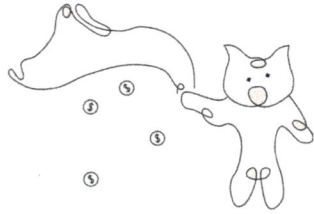

WHAT/WHEN

To engage in small, often trivial charges or expenses, typically in a way that feels petty or overly focused on minimizing costs.

> **"I thought we had booked an all-inclusive hotel, but they're nickel and diming us for everything—$2 for a beach towel!"**

AS A PRODUCT MANAGER

Controlling costs can feel like balancing on a high wire, especially when every penny matters. This idiom perfectly encapsulates the need to keep a sharp eye on finances without stifling innovation. In the tech industry, this could mean anything from negotiating the cost of server space to pinching pennies on marketing campaigns to ensure there's still budget left for user research.

One real-world example of a product manager "nickel and dime" scenario is when Netflix decided to tackle its password-sharing problem in 2023. The streaming giant realized too many viewers were piggybacking on friends' accounts, bleeding potential revenue. The product manager had to devise a plan to recover those lost dollars while keeping existing subscribers happy, which ended up being a new paid-sharing feature that nudged multiple households to pay their fair share.

For Product Managers, "nickel and dime" is less about being stingy and more about ensuring there's enough cash in the kitty to make something truly awesome. It's a constant dance—like telling the design team the gold-plated prototype has to wait, or reminding marketing that the budget for a fireworks-laden product launch might set off the finance team's alarm bells. Sometimes, it feels like you're the penny-pincher, the buzzkill, and the person digging under the office couch cushions for spare quarters all at once. But hey, nothing says "we made it work" like shipping a killer product on a shoestring budget… plus a few leftover nickels and dimes for pizza.

84. TAIL OFF

WHAT/WHEN

The period following the initial euphoria.

> **"Our production begins to tail off on Wednesday afternoon and is non-existent by Friday after lunch."**

AS A PRODUCT MANAGER

Sometimes, after a whirlwind of excitement, things just start to "tail off"—and spotting this trend early is essential for any savvy business. The idiom "tail off" perfectly captures the gradual decline in activity, interest, or performance that can quietly undermine even the most promising projects. In the dynamic world of startups, this might manifest as user sign-ups that initially skyrocket but then begin to level out without hitting the sustained growth targets.

A classic example of a product "tailing off" is the story of Google+. Launched with grand ambitions to rival Facebook, Google+ enjoyed a flurry of curiosity and initial sign-ups. However, over time, user engagement began to "tail off" as the platform struggled to offer compelling reasons for people to stay active. Google had to reassess its strategy, ultimately winding down the service as it failed to maintain the momentum against entrenched social media giants.

For Product Managers, watching metrics "tail off" can feel like throwing a fantastic party where the guests slowly start to disappear one by one—first the early birds, then the mid-enthusiasts, until you're left wondering why no one showed up for the after-party. It's about juggling declining user engagement while brainstorming new features that might reignite interest, all while keeping a straight face when the latest marketing push doesn't spark the revival you hoped for. Balancing optimism with strategic pivots, it's the art of turning the slow fade into a new crescendo before the spotlight completely dims (and no, Chad, adding more emojis to the homepage won't fix it).

85. TAILWINDS

WHAT/WHEN

Forces that complement the achievement of an objective.

> **"The test audience loved the film, and the studio agreed to the promotion budget. We've got the tailwinds behind us."**

AS A PRODUCT MANAGER

It's crucial to recognize when conditions are in your favor, especially if you want to propel your product (and your sanity) forward. This phrase perfectly captures the benefit of having external forces give you that extra push. In the tech industry, this could mean anything from an unexpected upswing in user adoption to a sudden hype wave around your product's core technology.

One real-world example of a product manager riding "Tailwinds" is Slack's meteoric growth during the rapid shift to remote work in 2020. As entire teams had to communicate virtually, Slack emerged as the go-to platform for streamlined messaging and collaboration. The product manager capitalized on this momentum by introducing features tailored for newly remote teams, effectively cementing Slack's place as the digital office water cooler.

For Product Managers, "Tailwinds" is less about heroism and more about letting the breeze carry you while you pretend you knew all along that the market would move in your favor. It's juggling sudden opportunities with a straight face—like discovering your platform went viral overnight right as someone on the team wants to tack on an experimental feature (spoiler: maybe next sprint, Chad). Sometimes, it feels like the wind is cheering you on while you're just trying to keep your hair from blowing straight into your eyes. But hey, no one ever complains about free momentum when it wafts in and makes the roadmap look like a stroke of genius.

86. HEADWINDS

WHAT/WHEN

Negative events that complicate or eliminate success.

> **"Joe says the budget was rejected, and Terry's not sure he wants to continue. With these headwinds, the project is dead."**

AS A PRODUCT MANAGER

Facing challenges is a natural part of any venture, especially when they threaten to blow you off course. This idiom perfectly encapsulates the idea that sometimes, despite our best-laid plans, we have to brace ourselves against forces beyond our control. In the tech industry, this could mean anything from contending with sudden market downturns to grappling with a competitor's surprise product launch.

One real-world example of a product manager facing "Headwinds" is when Netflix dramatically lost subscribers after increasing its subscription fees back in 2011. The sudden backlash and customer exodus were like unexpected gusts that threatened to derail the company's growth. The product manager had to quickly adapt the strategy—tightening budgets, exploring new content deals, and doubling down on streaming to stabilize the business.

For Product Managers, "headwinds" are less about heroism and more about not getting knocked off your feet while you're figuring out which way the wind is blowing in the first place. It's a balancing act with a brave face—like dealing with a late-breaking customer complaint while the CEO is asking for a roadmap overhaul (spoiler: no, Chad, that can't happen by tomorrow). Sometimes, it feels like you're the sailor, the lighthouse keeper, and the person who forgot to check the weather forecast all at once. But hey, nothing unites a team like the gust of urgent priorities and the adrenaline of steering through a storm.

87. CHEW ON SOMETHING

WHAT/WHEN

Taking the time to consider the elements of a decision.

> **"I'm not sure of the best way to proceed. I need to chew on it awhile."**

AS A PRODUCT MANAGER

Sometimes, taking a step back and giving yourself time to reflect is key to making a good decision. This is where the idiom "chew on something" comes in, suggesting that before jumping into action, it's worth giving the situation some serious thought. In a business setting, this can mean carefully considering a new product idea, evaluating customer feedback, or deciding on the next big move.

A great example of this occurred when Nike launched its "Nike Training Club" app. The company didn't rush to make it a major part of their business overnight. Instead, they took their time to "chew on" the concept, researching their target audience, testing the app with different features, and figuring out how to best integrate it into their overall product ecosystem. This approach allowed Nike to refine the app, adding features and improvements that better suited their users' needs, resulting in a successful and well-received product.

For Product Managers, "chewing on something" means not getting swept up in the whirlwind of constant pressure to act. It's about creating space to evaluate all angles and think through the possible outcomes, like when you're deciding whether to pivot a product feature after months of development. It's that moment when you sit down, grab a cup of coffee, and contemplate the 47 different ways your project could go wrong—before settling on the best path forward. It's not about procrastination, it's about being smart enough to know that the right decision is worth the wait. And who doesn't love a good think session before diving into another round of "what could possibly go wrong?"

88. CAN'T SEE THE FOREST FOR THE TREES

WHAT/WHEN

Focusing on details and missing the objective.

> **"Ford's Model T was the best-engineered car on the road, but people wanted colors besides black."**

AS A PRODUCT MANAGER

It's easy to get caught up in the small details when you're working on a big project. This idiom highlights the challenge of being too focused on individual parts and losing sight of the bigger picture. In business, it often happens when someone zooms in on one issue so much that they miss how it fits into the overall strategy or larger goal.

A prime example of this is when companies get stuck analyzing customer feedback on a single feature while ignoring broader market trends. Take, for instance, the time when Blockbuster was busy tweaking its DVD rental kiosks instead of seeing the massive shift to digital streaming. By focusing on one minor thing, they missed the huge change in the industry, leading to their downfall.

For Product Managers, "can't see the forest for the trees" often happens when you're deep in the weeds, overthinking a product tweak while forgetting the overarching vision. It's like obsessing over a button color that no one will notice while the entire user experience is in disarray. You're so busy making sure the tree next to you looks perfect that you forget to step back and see the forest full of other trees that also need attention. It's a classic case of thinking small while the world keeps moving in the other direction.

89. THROWING OUT THE BABY WITH THE BATHWATER

WHAT/WHEN

Failing to distinguish between the good and bad.

> **"When restructuring our product line, we must avoid throwing out the baby with the bathwater and protect the core features that customers value."**

AS A PRODUCT MANAGER

Sometimes, when you're solving a problem, it's easy to get so caught up in fixing the details that you end up throwing the baby out with the bathwater. This idiom highlights the danger of overreacting or going too far when attempting to resolve an issue, ultimately causing more harm than good. In the world of business, this can often happen when trying to streamline processes or cut costs without considering the full impact of those changes.

A prime example of "throwing out the baby with the bathwater" came when Gap rebranded its logo in 2010. In an effort to modernize its image, the company introduced a new logo, completely abandoning the iconic blue square. The backlash was immediate, with customers and designers alike criticizing the change as both uninspired and unnecessary. Gap quickly scrapped the new logo and returned to its original design, illustrating how a well-meaning move can alienate your audience if you're not careful.

For Product Managers, "throwing out the baby with the bathwater" serves as a reminder to assess what truly matters in a product and avoid drastic changes for the sake of innovation. It's like updating an app and accidentally removing features customers rely on (sorry, Chad, we *really* need offline mode). It's about finding a balance between fresh ideas and maintaining what makes your product work, because sometimes the things you think need fixing are the ones worth keeping the most.

90. FISH OR CUT BAIT

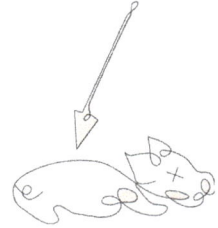

WHAT/WHEN

The pressure to decide.

> "We've looked at the problem up, down, and sideways. We have three choices. It's time to fish or cut bait."

AS A PRODUCT MANAGER

When it comes to making decisions, there's a moment where you just have to stop talking about it and take action. The idiom "fish or cut bait" perfectly sums up that point where indecision becomes a hindrance. It's about choosing whether to fully commit to something or walk away. In the fast-paced world of product management, this could be when you have to decide whether to push forward with a product feature that's on the fence or just drop it before the team wastes any more time.

Take for example the time when Microsoft decided to push ahead with the Windows Phone in the early 2010s. Despite being well behind competitors like Apple's iPhone and Google's Android, they kept adding features and trying to carve out a niche. But eventually, they had to face reality: either commit big time and improve their ecosystem or cut bait and move on. The decision to cut bait came in 2014 when Microsoft announced they were no longer making their own phones, letting go of the platform altogether.

For product managers, "fish or cut bait" is about making the call when things get murky. It's about resisting the temptation to endlessly analyze and instead accepting that sometimes the only way to know if a decision will pay off is to dive in or pull out. And just like fishing, you can't just keep casting the line into the same spot hoping for a different result—eventually, you've got to either reel it in or move on to the next opportunity. It's a constant balance between risk and reward, kind of like watching your inbox fill up while deciding whether to send that risky feature update at 4:59 PM on a Friday.

91. SWING FOR THE FENCES

WHAT/WHEN

Make a major effort.

> "We don't want to be another also-ran. This is the time we need to swing for the fences."

AS A PRODUCT MANAGER

When you go all out on a project, sometimes you just need to "swing for the fences." This idiom is all about taking a big, bold risk in the hopes of hitting a home run. In the world of business, it's that moment when you decide to go for the grand idea that could either elevate your brand or make you look like you missed the ball entirely. It's daring, it's risky, and, above all, it's the stuff of legends—if it works.

Consider a real-world example with Elon Musk and the launch of the Tesla Roadster into space. Musk and SpaceX took the audacious step of launching a car into orbit around Earth just for the heck of it. Did they know it would work perfectly? Absolutely not. But they swung for the fences, aiming to make a statement about pushing boundaries and redefining what's possible. And, shockingly, it actually worked—creating headlines and a ton of buzz for Tesla in the process.

For product managers, "swinging for the fences" is about taking the boldest step, even when the outcome is uncertain. It's the tightrope walk of risking it all—whether it's pushing for a groundbreaking feature launch or backing a product that's a bit too "out there" for most. It's a gamble, but sometimes, you have to bet big to win big. Just remember, if you miss, it's not failure; it's just one more thing to laugh about with your team over beers while planning the next big swing.

92. HITTING THE HIGH SPOTS

WHAT/WHEN

Emphasizing the major details.

> **"I don't have time for a long, confusing message. Just hit the high spots."**

AS A PRODUCT MANAGER

Sometimes in life, you just need to focus on the big wins and leave the little stuff behind. This idiom is all about prioritizing what truly matters and making sure you're hitting the highlights rather than getting bogged down in every minor detail. In the business world, this could mean zeroing in on major product features that will drive sales while letting smaller bugs and tasks take a backseat for the moment.

A perfect example of "Hitting the High Spots" is how Elon Musk and his team managed the early years of Tesla. Rather than getting caught up in the minutiae, they focused on perfecting the groundbreaking Model S. By zooming in on that one key vehicle, they were able to elevate the entire brand and establish Tesla as the future of electric vehicles—while, let's be honest, the whole "affordable mass-market car" thing took a bit longer to get right.

For Product Managers, "hitting the high spots" is less about checking off every box on a to-do list and more about knowing where to focus your energy. It's about choosing the big wins—the features that will move the needle—while letting the less exciting tasks hang out in the background (sorry, spreadsheet cleanup). It's like planning a road trip and knowing you're going to skip the random roadside attractions in favor of hitting the Grand Canyon. Sure, the roadside attractions have their charm, but in the end, you're after the big moments that will make your journey worth it. Meanwhile, Chad's over there trying to fit in every tourist trap because, you know, *all* the sights matter. Chad, we're just trying to get to the Grand Canyon, not collect 100 magnets, okay?

93. FOCUS ON THE BIG PICTURE

WHAT/WHEN

The solution to fussing over details.

> **"We can't spend all day fighting over the font type in the article. We need to focus on the big picture, the content, not its look."**

AS A PRODUCT MANAGER

Being able to focus on the big picture is crucial when trying to achieve long-term goals, especially when day-to-day details can often derail progress. This idiom captures the idea of not getting bogged down in the minutiae but rather stepping back and seeing how everything fits together. In the world of business, it can be easy to lose sight of the larger strategy when you're constantly addressing immediate concerns or distractions.

A classic example of focusing on the big picture can be seen in Elon Musk's approach with SpaceX. Despite the constant technical setbacks and failed launches early on, Musk kept his eyes on the ultimate goal of making space travel more affordable and sustainable. It wasn't just about fixing individual rockets—it was about getting humanity to Mars. He didn't let the day-to-day frustrations distract him from the end game, which, as it turns out, was a pretty big deal.

For Product Managers, "focusing on the big picture" often means resisting the temptation to dive into every tiny issue that arises. It's about saying "no" to every little tweak that sounds important and remembering that you're building something bigger than just the next release. Sometimes, it feels like everyone's trying to pull you into a rabbit hole—bug fixes, feature requests, endless meetings—but you've got to stay strong, like a general overlooking the battlefield while the team scrambles over the trenches. The challenge is knowing when to zoom in and when to keep your eye on the horizon. It's a delicate dance, but it's the only way to get somewhere instead of just spinning your wheels in the weeds.

94. SPENDING DOLLARS TO SAVE DIMES

WHAT/WHEN

Targeting the wrong objectives.

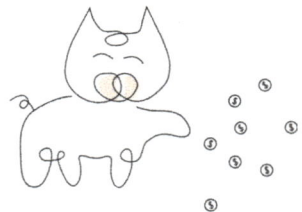

> **"We don't need to develop proprietary software when off-the-shelf serves the purpose. We'll be spending dollars to save dimes."**

AS A PRODUCT MANAGER

When it comes to managing budgets, the idea of "Spending Dollars to Save Dimes" can be a real head-scratcher. This idiom highlights the paradox of making a large investment in order to save a small amount in the long run. It's like buying a brand-new sports car to save on gas for your 10-mile commute, which, spoiler alert, doesn't make much sense.

A classic example of this idiom in action can be seen with the Juicero, a $400 juice press that was supposed to revolutionize the way we drink juice. The machine required specially designed juice packs, and after an investigation, it was revealed that you could simply squeeze the juice packs by hand for the same result. Essentially, you were paying hundreds of dollars for a device that did something you could do with your own two hands—saving you exactly nothing. It's the epitome of "spending dollars to save dimes," where a well-intentioned product ends up costing far more than it ever saves.

For Product Managers, it's all about knowing when to invest in innovation and when to keep it simple. Sometimes, you pour a lot of resources into a fancy solution only to realize it's not solving a real problem—just creating a bigger one. It's like building an elaborate, expensive system to fix something that could have been solved with a simpler tool. It can feel a little like buying the high-end kitchen appliance that promises to chop, slice, and dice, only to end up using a regular knife because the new gadget is too complicated to be worth the trouble.

95. GETTING LOST IN THE WEEDS

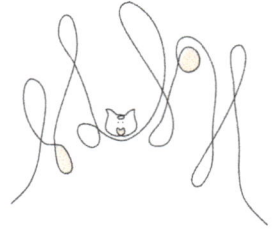

WHAT/WHEN

The over-emphasis on details.

"We're spending too much time and money on this landing page. We're getting lost in the weeds."

AS A PRODUCT MANAGER

Getting lost in the weeds happens when you focus on the small details so much that you forget about the bigger picture. It's like being so wrapped up in the color of the napkins for an event that you forget to actually send out the invitations. This idiom perfectly captures the tendency to get sidetracked and lose track of what truly matters. In any job, especially in fast-paced industries, getting caught in the weeds can be a huge time-sink and a productivity killer.

A classic example of "getting lost in the weeds" is when a product manager spends hours refining the wording of a button on the app instead of making sure the app actually works. Take Google's failed attempt at releasing Google Glass—while the team was obsessing over the design and features, the underlying technical issues weren't addressed. By the time they realized the core product needed more attention, they were knee-deep in the weeds of design and user experience tweaks that were ultimately irrelevant.

For Product Managers, getting lost in the weeds is like staring at a single pixel of a website for so long that you forget you're building an entire platform. It's chasing down every minor detail, like whether the font size is exactly right, while missing the glaringly obvious flaws, like, say, the app crashing every five minutes. It's easy to fall into the trap of polishing every little aspect, but sometimes you just need to step back and remember that no one cares about the font size when the whole page is broken. And yet, there's Chad, asking if we can change the color of the submit button one more time—no, Chad, we can't, because we're still fixing the fact that the app won't load at all.

96. BELLS AND WHISTLES

WHAT/WHEN

Product features that attract consumer interest.

> "We have a good product, but nothing stands out from the competition. We need to add some bells and whistles."

AS A PRODUCT MANAGER

When it comes to product development, it's easy to get caught up in adding all the bells and whistles, thinking that more is always better. This idiom describes the tendency to overload a product with extra features, often unnecessary or overly complex, in an attempt to make it stand out. In reality, this can lead to overcomplication, leaving users confused and potentially frustrated.

One memorable example of "bells and whistles" gone wrong happened with Samsung's Galaxy Note 7. The phone was packed with every possible feature, including an innovative stylus, an edge screen, and more, but all those shiny extras didn't matter when the phone's battery would overheat and catch fire. Despite all the high-tech features, it became infamous for the wrong reasons, showing that sometimes less is more—and some features are better left off.

For Product Managers, chasing after every "bell and whistle" is a tempting but dangerous game. It's like trying to fill a swimming pool with every toy imaginable, only to realize you've just created a water park that nobody actually wants to visit. The key is to focus on what truly benefits the user rather than getting distracted by flashy features that make great demos but lack real value. It's a fine balance between innovating and resisting the urge to toss in every feature under the sun, hoping to make the product seem more impressive than it really is.

97. DROWNING IN DETAIL

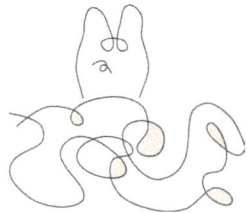

WHAT/WHEN

When information is overwhelmed by too much data.

> "I bought a software program to edit videos. I never learned to use it because the instructions were page after page of jargon, drowning me in details."

AS A PRODUCT MANAGER

When you're trying to make progress on a project, it's easy to get bogged down in the minutiae. This idiom describes the all-too-common situation where you're so deep in the weeds that you lose sight of the bigger picture. In a product development context, this could be spending hours debating over font sizes or color palettes when you're really just trying to finalize a major feature for launch.

A prime example of "Drowning in Detail" comes from an infamous design review at a major tech company. The team spent days obsessing over the alignment of icons and the shade of blue for the app's background, while the actual functionality of the app was slipping through the cracks. At some point, the product manager had to step in, push for clarity, and remind the team that the app's core features—like, you know, actually working—were far more important than how the button hovered over the screen.

For Product Managers, "Drowning in Detail" often feels like you're spending an eternity debating something that, in the grand scheme of things, won't matter at all. It's like putting a magnifying glass over a speck of dust while the entire house is on fire. It can be hard to pull back and see the forest for the trees, especially when the trees are looking *so* darn perfect. But hey, sometimes a good old reality check and a well-timed "Let's move on" are the only life preservers you'll get in that sea of specs and spreadsheets.

98. TOUCH BASE

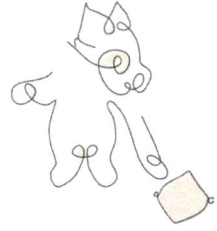

WHAT/WHEN

A short, informal meeting or communication.

> "My neighbor works in the purchasing department of GM. I'll touch base with him for ideas to offer our products to them."

AS A PRODUCT MANAGER

Sometimes, staying in the loop is as important as solving the problem itself. The phrase "touch base" embodies that moment when you just need to make sure everyone is on the same page, even if you're not diving deep into the details. It's about keeping the communication lines open, whether it's a quick check-in or a brief update to ensure everything's moving forward smoothly.

A great example of "touching base" in the corporate world comes from the 2008 financial crisis when companies were scrambling to find out how their assets were really holding up. Executives had to regularly "touch base" with their teams to get fast, consistent updates on what was happening in different departments. These check-ins ensured that no one was left in the dark while they worked through the rapidly changing landscape.

For product managers, "touching base" is like making sure no one accidentally launches an ad campaign with a glaring typo or misses an important deadline—before it becomes an emergency. It's a moment to quickly align everyone, clear up any misunderstandings, and reassure the team that everything's under control. It might feel mundane at times, but it's the secret ingredient to making sure everyone is moving in the same direction, even if it's just a quick reminder to double-check that we are not overspending on Google Ads (yes, Chad, I'm talking to you).

99. TOUCHPOINTS

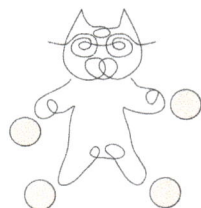

WHAT/WHEN

Points of contact between a business and a customer.

> "We need to nourish our relationships with customers. Our touchpoints are the times they hear, read, or touch something that carries our brand."

AS A PRODUCT MANAGER

When you're managing a product, it's essential to be aware of all the key interactions with customers and stakeholders. These are your "touchpoints"—the moments where your product connects with people and leaves an impression. Touchpoints can come in many forms, like a user interacting with a mobile app, getting an email notification, or even encountering a problem during checkout. It's all about ensuring that every interaction is as smooth and memorable as possible.

A prime example of "touchpoints" in action is the customer experience strategy at Starbucks. Their app, which allows users to order ahead, collect loyalty points, and customize their drinks, is filled with touchpoints that keep customers engaged. Whether it's a personalized offer, an easy-to-navigate interface, or a friendly push notification reminding you it's time for a coffee, Starbucks has fine-tuned these moments to build a loyal customer base.

For product managers, "touchpoints" are less about checking a list of interactions and more about crafting an experience that makes every user feel like a VIP. It's not just about fixing problems; it's about making sure users remember that perfect login screen, that time the app made them smile, or the moment they realized they didn't have to call customer support because everything was so intuitive. Sometimes, it's about getting those moments just right—like perfecting the way your app says "hello" so your users feel welcomed before they even touch the screen.

100. CASH COW

WHAT/WHEN

A business or product that delivers steady revenues without significant effort.

> "Our website hosting service is almost automated, with very little intervention required. It's become a cash cow."

AS A PRODUCT MANAGER

A cash cow can be a real moneymaker in any business—if you know how to spot it. This idiom refers to a product, service, or business that reliably generates a steady stream of income, often with minimal ongoing effort or maintenance. In the retail world, for example, a popular, evergreen product can continue to rake in profits year after year without much need for reinvention.

Take Coca-Cola, for example. Despite the rise of trendy new beverages, Coca-Cola has remained a staple in refrigerators across the globe. The iconic drink has been a cash cow for the company for decades, providing a reliable revenue stream while the company explores new product lines. Even as they expand into bottled waters and energy drinks, Coca-Cola still knows its golden ticket lies in that classic, fizzy formula.

For Product Managers, a "cash cow" is less about constant innovation and more about making sure you don't ruin a good thing while riding the gravy train. It's about ensuring the product runs smoothly, maintaining customer loyalty, and occasionally dusting it off with a fresh marketing campaign. You're not trying to reinvent the wheel, you're just making sure it doesn't fall off the cart while you're busy working on the next big thing (even if the next big thing isn't a sure bet). Think of it as a sweet, steady gig—like being the person who keeps the vending machine stocked with snacks everyone loves but is too tired to actually appreciate.

101. SWEET SPOT

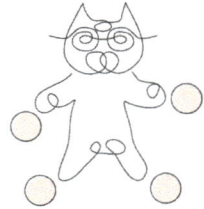

The perfect balance between opposing forces.

> **"We can't compete with Walmart on price. We need to find the sweet spot between low price and customer service."**

AS A PRODUCT MANAGER

Finding the "sweet spot" is like hitting the bullseye on a dartboard—it's that perfect balance where everything just comes together. This idiom captures the moment when all the factors align, creating an ideal situation with just the right amount of effort, timing, and resources. It's like finding the perfect spot to park on a crowded street—convenient, accessible, and with just enough space.

A great example of hitting the "sweet spot" is Costco's approach to its membership model. Costco figured out the perfect balance between offering quality products at bulk prices while still ensuring they could turn a profit. By keeping their prices low and membership fees high, they hit the sweet spot between value for customers and maintaining strong margins for the business. Their success came from balancing the wants of budget-conscious shoppers with the need for cost-effective inventory management.

For Product Managers, finding the "sweet spot" is about creating something that checks all the boxes—meeting customer needs, being scalable, and making a profit. It's like Costco offering a 36-pack of paper towels; it's big enough to save money, but not so excessive that it feels wasteful. When you hit that sweet spot, it's like finding the perfect product mix that keeps everyone happy, from the customers to the bottom line.

102. HIT THE GROUND RUNNING

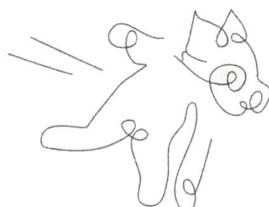

WHAT/WHEN

Complete preparation before execution.

> **"The new ecommerce App will be released before Thanksgiving. We need to hit the ground running with our exclusive App-only promotion."**

AS A PRODUCT MANAGER

You've probably heard the phrase "hit the ground running" before, but what does it actually mean? It's all about jumping into action without hesitation, ready to tackle whatever challenges come your way. This idiom is especially useful in high-pressure environments where you need to act quickly and deliver results, like when you've got a new product launching and the clock is ticking.

A great example of "hitting the ground running" comes from Under Armour's launch of its innovative performance gear. When the company first introduced its moisture-wicking fabric in the early 2000s, it had to quickly capture attention in a market dominated by giants like Nike and Adidas. Under Armour didn't waste time—leveraging aggressive marketing, celebrity endorsements, and a fresh product line, they successfully disrupted the industry and built a strong presence almost overnight.

For Product Managers, "hitting the ground running" is about diving in headfirst with little time for hand-holding. You're expected to absorb the chaos, make decisions fast, and keep everyone moving forward—no matter how many unexpected hurdles appear. It's like being a sprinter, launching out of the gate full speed, while keeping track of your competitors and making sure your shoes don't trip you up. At Under Armour, they'd tell you it's not just about staying in the race—it's about crushing the competition from the get-go.

103. BEST THING SINCE SLICED BREAD

WHAT/WHEN

A revolutionary, successful new product.

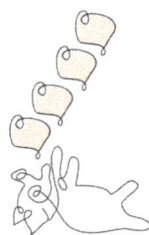

> "People are always reluctant to try new things. We need to show them this product is the best thing since sliced bread."

AS A PRODUCT MANAGER

When you think about taxes, the last thing that comes to mind is convenience. But sometimes, a new idea comes along that makes a tedious task feel almost... easy. This idiom perfectly captures that moment when a solution arrives that transforms a complicated process into something so simple that it feels like the best thing since sliced bread. In the world of finance, this could be the launch of TurboTax.

One real-world example of this is TurboTax, which revolutionized the way people file their taxes. Before its arrival, filing taxes was often a confusing, time-consuming task that required a professional or hours of frustration. TurboTax introduced an intuitive, step-by-step process that walked users through their tax filings with ease, turning a dreaded annual chore into something much more manageable. It wasn't just another software tool; it was a game-changer in personal finance.

For Product Managers, creating something that feels like the "best thing since sliced bread" is about taking a process that seems impossible and making it simple. It's like TurboTax, where you're not just designing a product; you're creating an experience that takes the headache out of something most people avoid. Of course, you have to deal with endless updates to accommodate changing tax laws and user complaints about refund times, but when it all clicks, it's like giving your customers the gift of time and sanity—two things we could all use more of.

104. EARLY BIRD GETS THE WORM

WHAT/WHEN

An appeal to quick action, the incentive to be the "first mover."

> "I hear ABC Co is ready to launch their tax software. We need to be in the market first, as the early bird gets the worm."

AS A PRODUCT MANAGER

Waking up early might seem like a chore, but in some cases, it's the secret sauce to success. The idiom "the early bird gets the worm" speaks to the benefits of being proactive and getting a head start before the competition. In the world of media and content creation, this is particularly true. Whether it's being the first to cover a trending topic or getting ahead of the curve with new content, those who act first often reap the rewards.

One prime example of this comes from Joe Rogan, whose early decision to dive into podcasting transformed him into a global media mogul. While many were still skeptical about the medium, Rogan recognized the potential early on and built a loyal audience. By being one of the first to establish himself in the podcasting world, he locked in a massive following, which only grew as the platform gained popularity. His deal with Spotify in 2020 for $100 million wasn't just about being famous—it was about recognizing a trend and being there when it mattered.

For Product Managers, "the early bird gets the worm" is a reminder that success often favors those who take the initiative. It's about acting before your competition does, even if it means getting up earlier than you'd like. Just like Joe Rogan wasn't waiting for someone else to make podcasting cool, you can't wait around for a perfect moment to launch your product. By getting ahead of the curve, you put yourself in a position to grab the worm before anyone else even realizes it's there.

105. TOO GOOD TO BE TRUE

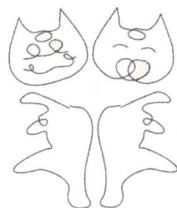

WHAT/WHEN

Favorable consequences are likely to have a hidden condition.

> **"Roger offered a deal that seems too good to be true. He knows something we don't. We need to be careful."**

AS A PRODUCT MANAGER

Sometimes, things seem just *too good to be true*—and in the case of Elizabeth Holmes and Theranos, that was an understatement. This idiom perfectly captures the disbelief we feel when something sounds amazing but feels a little off. It's the gut feeling that tells you to proceed with caution when everything appears perfect on the surface, especially in business or tech.

Take Theranos, for example. The company promised to revolutionize healthcare with a small device that could perform hundreds of tests using just a drop of blood. The idea was so enticing that investors, doctors, and media outlets jumped on board, thinking they were witnessing the future of medicine. But, of course, it turned out to be "too good to be true"—the technology didn't work as promised, and the company was exposed for misleading investors and patients. The result? A massive scandal that shook Silicon Valley to its core.

For Product Managers, the Theranos debacle serves as a cautionary tale of how easy it is to get caught up in a hype machine. Launching a product with the promise of groundbreaking innovation is exciting, but managing those expectations is crucial. You can't promise a miracle when you're still figuring out the basics. It's like telling your customers that your new app will make them breakfast, only to find out you forgot to include the timer. So, before you promise the moon, remember: sometimes "too good to be true" isn't a challenge, it's a red flag.

106. SECOND MOUSE GETS THE CHEESE

WHAT/WHEN

The advantage of learning from the first mover.

> "Facebook dominates the social media industry, but we can learn from their mistakes and offer a better product."

AS A PRODUCT MANAGER

The idiom "Second mouse gets the cheese" perfectly illustrates the power of patience and strategic timing. Rather than rushing in when the competition is at its peak, it's about waiting for the right moment, learning from the early movers' mistakes, and then swooping in with a refined and more successful approach.

A great example of this is Ring, the home security company. While Ring wasn't the first to introduce a smart doorbell, it entered the market after competitors like SkyBell and Zmodo had already launched similar products. However, Ring's approach was more polished—offering superior marketing, integration with other smart home devices, and a more user-friendly experience. Ring took the lessons learned from the early adopters, addressed their shortcomings, and gained significant traction. In 2018, Amazon acquired Ring for nearly $1 billion, demonstrating how the second mouse can grab the cheese by learning from others' mistakes and offering a better solution.

For product managers, this is a perfect example of how timing and observation can lead to success. It's not always about being first to market, but rather about understanding what the market really needs and executing with precision. Ring didn't just launch a smart doorbell; they launched an ecosystem that solved real customer pain points, positioning themselves to dominate the home security market. The second mouse doesn't just get the cheese—they redefine the game.

107. GO VIRAL

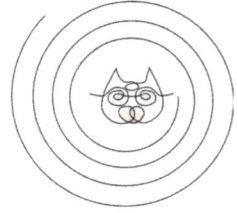

WHAT/WHEN

The phenomena of the geometric spread of sales by consumer word of mouth.

> **"We need to develop a campaign that will sustain itself. We need to go viral."**

AS A PRODUCT MANAGER

It's no secret that everyone wants to become an overnight sensation, and with the right conditions, it can happen in the blink of an eye. The phrase "go viral" captures this perfectly, describing the sudden, explosive spread of content across the internet. In the world of social media, this could mean anything from a funny cat video to a dance challenge that takes over every platform.

A great example of something "going viral" in the wild world of the internet is the "Ice Bucket Challenge" from 2014. What started as a simple challenge to raise awareness for ALS quickly turned into a global phenomenon, with celebrities, athletes, and everyday people dousing themselves in freezing water. It spread like wildfire, raising millions for charity and becoming a cultural moment no one could avoid.

For product managers, "going viral" isn't necessarily about creating the next big meme but about capitalizing on unexpected opportunities. It's like launching a product feature that catches the internet's attention—except instead of being able to enjoy the moment, you're left scrambling to scale your infrastructure because your servers weren't prepared for a million new users (sorry, Chad, that "optimization" thing we talked about is a priority now). So, while going viral can feel like a blessing, it's also a reminder that with great popularity comes great responsibility, and sometimes, a lot of very nervous clicks on the refresh button.

108. DEVIL IN THE DETAILS

WHAT/WHEN

The importance of details and their effect on outcomes.

> **"Rolling out a new software update seems simple, but the devil is in the details."**

AS A PRODUCT MANAGER

You've probably heard the phrase "the devil is in the details," and while it sounds ominous, it's actually a reminder that overlooking the small stuff can cause huge problems down the line. This idiom perfectly captures how the tiniest mistakes, when left unchecked, can snowball into major setbacks. Whether you're launching a new product, negotiating a deal, or making sure that everything in your office space is running smoothly, those little details can be the difference between success and failure.

A real-world example of the "devil in the details" happened with WeWork's rapid rise and fall. The coworking giant, once valued at $47 billion, was eventually brought down by a series of overlooked details—most notably, its business model. WeWork's financials were filled with inconsistencies, like inflated revenue projections, questionable spending habits, and even the fact that founder Adam Neumann was using company funds to purchase real estate. All these seemingly small details added up and, when investors started digging deeper, the company's billion-dollar valuation came crashing down. What seemed like minor oversights at first turned into a catastrophic downfall.

For Product Managers, "the devil is in the details" means paying attention to everything, no matter how trivial it seems at first. It's like trying to make sure your team's project plan is airtight, only to realize you missed the small clause in the contract that could void your entire partnership. It's exhausting when the product launch is going smoothly, only to realize the website's signup form is broken, or the latest update didn't actually improve the app's performance. But catching these details early is what keeps you from becoming the next cautionary tale of a high-flying company that crashed because someone forgot to check the fine print.

109. ROTTEN APPLE

WHAT/WHEN

The influence of one negative member can destroy a team.

AS A PRODUCT MANAGER

The phrase "rotten apple" is often used to describe someone who's the problem in a group or situation. It's like that one person who brings the whole team down without even trying, leaving a trail of bad decisions behind them. In the workplace, this could refer to someone who consistently undermines the team's progress or creates unnecessary drama, making everyone else's job harder.

A perfect example of a "rotten apple" in the business world comes from the infamous 2017 incident with the leadership team at Uber. Travis Kalanick, the company's CEO at the time, was seen as a "rotten apple" due to his controversial management style, which included fostering a toxic culture and making decisions that ultimately hurt the company's reputation. His leadership led to internal chaos, forcing him to step down and leave a power vacuum that took years to recover from.

For Product Managers, dealing with a "rotten apple" can feel like trying to save a dish that's already spoiled. It's about making tough decisions and separating the spoiled from the good. Whether it's addressing toxic behaviors, handling impossible expectations, or navigating conflicts that derail progress, you have to stay cool while your team's morale is slowly being poisoned. Sometimes, you're the one handing out the tough love, but other times, you just have to remove the rotten apple from the equation altogether before it ruins the whole bunch.

110. EAGER BEAVER

WHAT/WHEN

A description of a person eager to begin and complete a task.

"Michael's an eager beaver. We need to be sure we don't let him get over-extended."

AS A PRODUCT MANAGER

When you're in the fast-paced world of product management, it's important to have a team member who is always ready to go the extra mile, especially when deadlines are tight. This is where the phrase "eager beaver" comes in. It's used to describe someone who's so enthusiastic and motivated that they're often the first to jump in, no matter how big the challenge. For a product manager, having an eager beaver on the team is like having a turbocharged teammate who's always asking, "What's next?"

A prime example of an "eager beaver" in action could be Steve Jobs during the development of the original iPhone. Jobs was famously relentless in his pursuit of perfection and would often push his team beyond their limits. He'd demand the impossible and keep them on their toes, always motivating them to work harder, innovate faster, and think bigger. The team may have rolled their eyes at times, but that unshakable drive to innovate was exactly what pushed Apple to revolutionize the smartphone industry.

For Product Managers, being an "eager beaver" is more about enthusiasm than perfection. It's like having someone in the room who raises their hand to volunteer for every task, even the ones that make everyone else roll their eyes. Sure, they might need a little guidance on pacing and maybe a few reality checks, but that over-the-top energy is contagious. In the end, the eager beaver is the first to take the initiative, which—while occasionally leading to chaos—is exactly what you need when the project needs that extra push to cross the finish line. And if you're lucky, they'll even bring coffee for everyone when it gets really late.

111. DEEP DIVE

WHAT/WHEN

An in-depth examination of a certain topic, issue, or subject.

> **"Before we decide on the approach, we need a deep dive into the customer demographics."**

AS A PRODUCT MANAGER

It's crucial to get into the nitty-gritty of a situation when you're trying to figure out how to solve a big problem. This idiom perfectly captures the idea of going beyond surface-level understanding and diving into the complexities of an issue. Whether it's troubleshooting a major system glitch or analyzing user feedback to redesign a feature, taking a "deep dive" means you're not just skimming the surface; you're exploring every corner to find the root cause.

Take, for example, when Netflix decided to overhaul their recommendation algorithm to improve user experience. They didn't just tweak a few settings and call it a day. They dug deep into user data, explored patterns in viewing habits, and even considered what users weren't watching. The result? A more personalized experience that kept subscribers coming back for more content. It was a deep dive that ultimately paid off in big subscriber growth.

For Product Managers, the "deep dive" is all about getting your hands dirty in the details, even if it means wading through a mountain of data or sitting through hours of feedback sessions. It's the kind of dive where you might start out looking for one answer and end up uncovering a treasure trove of unexpected insights. It's like diving into a pool, but you're pretty sure someone just threw in an octopus and a scuba tank full of mystery. The depth of your analysis might get a little overwhelming, but at least you're guaranteed to surface with something useful—unless you end up being stuck with the octopus, of course.

112. DRAIN THE SWAMP

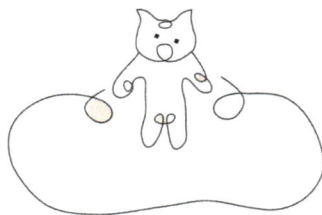

WHAT/WHEN

The process of eliminating an old culture or process in preparation of creating a new one.

> **"If we're going to revitalize the company, we need to drain the swamp first."**

AS A PRODUCT MANAGER

When you're trying to improve an organization or fix a broken system, sometimes you need to take a step back and get rid of the things that are weighing you down. This idiom perfectly captures the essence of cleaning up the mess and starting fresh. In politics, business, or even personal life, "draining the swamp" implies clearing out corruption, inefficiencies, or anything that stands in the way of progress.

A real-world example of "draining the swamp" can be seen in the early days of Amazon when Jeff Bezos made the decision to refocus on the company's core mission. At one point, the company was spread too thin with a variety of ventures that weren't necessarily contributing to their long-term success. By cutting the dead weight and streamlining operations, Amazon could focus on what it did best—selling books online. This "drain the swamp" moment allowed Amazon to evolve into the global e-commerce giant it is today.

For Product Managers, "draining the swamp" is about clearing away the unnecessary clutter so your product can thrive. It's like when you're working on a new feature, and you realize that half the feedback you've been getting is from people who are not even signed-in active users (thanks, Chad, again). Sometimes, you have to make tough calls, remove the distractions, and take out the things that don't add value. It's less about dramatic change and more about creating the space for something better to grow. And let's be honest, the real trick is doing it without getting bogged down in the swampy politics of office drama and email chains that never end.

113. GOING WHOLE HOG

WHAT/WHEN

Committing to an outcome without reservations.

> "We're in a race between getting funded or going bankrupt. We must go whole hog into program to have a chance."

AS A PRODUCT MANAGER

When you decide to "go whole hog," it means you're committing fully to something with no reservations, no half-measures, just a full, unfiltered dive in. This idiom is all about being bold and all-in, no holding back. Whether it's investing your time and energy into a big project or making a risky decision, you're putting all your chips on the table. In the tech world, this could be the equivalent of making a major shift in a company's product strategy, like Google diving headfirst into the smartphone market with the Pixel.

Take Sundar Pichai's leadership at Google as an example of "going whole hog." In 2014, when he took over the helm as CEO, Pichai fully committed to expanding Google's hardware business, including the ambitious launch of the Pixel phone. Despite the risks in competing against tech giants like Apple and Samsung, Pichai didn't hold back. He made Google's foray into hardware a core part of the company's strategy. The Pixel phone, while not a massive instant hit, represented Pichai's all-in approach to pushing Google into new territory, and over time, it's helped strengthen Google's brand beyond just search and software.

For Product Managers, "going whole hog" means putting everything into a product or project with full confidence, no looking back. It's about backing a risky decision because you believe in the potential, even when others might doubt it. And sometimes, it feels like riding a rollercoaster where you can't stop, even if you want to. But when it clicks? You feel like you've just nailed the perfect pitch. It's the kind of commitment that may lead to some bumps along the way, but it also comes with the satisfaction of knowing you gave it your all—just like Sundar Pichai did with Google's hardware push.

114. STRIKE WHILE THE IRON IS HOT

WHAT/WHEN

Reinforcing or exploiting an advantage.

> "Techcrunch just named our video software one of the Ten Best Products of the Year. We need to strike while the iron is hot."

AS A PRODUCT MANAGER

When opportunities present themselves, it's crucial to act swiftly—before the moment passes. "Strike while the iron is hot" perfectly captures the essence of taking immediate action when conditions are right. In the fast-paced world of business, this could mean jumping on a trend that's about to explode or rolling out a product at the perfect time when demand peaks.

One classic example of striking while the iron is hot is Zoom during the COVID-19 pandemic. In early 2020, when the world suddenly transitioned to remote work and virtual meetings, Zoom capitalized on the growing demand for a reliable video conferencing platform. As businesses, schools, and families scrambled to connect virtually, Zoom quickly became the go-to tool, skyrocketing in popularity and cementing its place as an essential part of daily life.

For Product Managers, striking while the iron is hot can feel like recognizing when a shift in the market is happening and acting before it's too late. It's about moving fast when you know your product is primed for success—just like Zoom did by seizing the remote work boom. Of course, there's a risk of the market cooling down or a new competitor swooping in, but when you hit that sweet spot, it feels like you've unlocked a cheat code. If you wait too long, though, you're just staring at a cold iron wondering where your moment went.

115. WHISTLEBLOWER

WHAT/WHEN

A term describing someone who makes organizational secrets public.

> **"I'm worried that we exaggerated the results of the test. It will blow up in our face with a single whistleblower."**

AS A PRODUCT MANAGER

It's essential to speak up when you witness something wrong, especially when it has the potential to affect millions of people. This idiom speaks to the courage required to bring hidden truths to light, often at great personal risk. In the tech world, this could mean exposing harmful practices or unsafe products, even when doing so puts you at odds with your employer or industry norms.

One of the most famous whistleblowers of our time is Edward Snowden, who revealed classified information about NSA surveillance programs. However, another powerful example is Frances Haugen, a former Facebook product manager who came forward in 2021 with internal documents that showed how Facebook's algorithms were amplifying harmful content, contributing to misinformation, and negatively impacting the mental health of young users. Haugen's decision to disclose this information was a pivotal moment in tech transparency, as it forced both the public and lawmakers to confront the ethical implications of social media's impact on society.

For Product Managers, being a "whistleblower" is less about being a martyr and more about sounding the alarm when you see something that could harm users or undermine the product's integrity. It's about having the integrity to stand up and say, "Hey, this isn't right," even when it feels like you're the only one willing to do so. In Haugen's case, it meant going against the very company she had once worked for, risking her career and reputation to bring attention to the dangers Facebook was enabling. It's a reminder that sometimes, being a Product Manager means protecting people from the things they don't know are happening behind the scenes, even if it's uncomfortable or unpopular. But hey, if you're going to risk it all, at least you can sleep knowing you did the right thing—just like Haugen did, despite the fallout.

116. BIG FISH IN A SMALL POND

WHAT/WHEN

A person or organization considered influential in a limited scope.

> **"The mayor of our town may be a big fish in the community but has little clout beyond the city limits."**

AS A PRODUCT MANAGER

It's easy to think that being the top dog in a small pond makes you a big deal. This idiom is a reminder that sometimes, what seems like greatness is actually just being the largest fish in a very small, very shallow pool. In many cases, we can all relate to the idea of being the big fish, whether in the workplace, a social circle, or even within a small project team.

For example, in the world of startups, a founder may feel like a genius when they're the only one calling the shots. This happened with companies like the early days of Google, where Larry Page and Sergey Brin were the stars of their small, scrappy startup. As the company grew, their roles expanded, but so did the competition, and suddenly, they weren't the biggest fish anymore—they were just a couple of very influential ones swimming in a much larger pond.

For product managers, being a "big fish in a small pond" often feels like being the one who knows the most in a niche area. It's a small victory—until you realize that your pond is so small it's basically a puddle, and there's a tidal wave of new problems coming. It's the moment when you're leading a small feature rollout, and everything feels like it's under control, only to be hit by the reality of a much larger product launch down the line. You may have been the biggest fish before, but suddenly, you're just a small fry in a much larger ocean of challenges, and good luck with that!

117. NUMBER CRUNCHER

WHAT/WHEN

A person whose primary purpose is to record and analyze data.

> **"If Bill weren't my number cruncher, I would be lost. He can make sense of whole columns of confusing, conflicting data."**

AS A PRODUCT MANAGER

When you're dealing with a flood of numbers and trying to make sense of them, you need a number cruncher who can break down the data into something more digestible. This phrase refers to someone who has the rare ability to analyze and interpret vast amounts of information quickly and accurately, often in high-pressure situations. Whether it's calculating financial ratios or forecasting trends, the number cruncher is there to ensure that no detail slips through the cracks. In the world of business, the ability to turn raw data into actionable insights is crucial for making informed, strategic decisions.

A prime example of a number cruncher in action is Charlie Munger, the longtime partner of Warren Buffett at Berkshire Hathaway. Munger is known for his sharp analytical mind, using data and logical reasoning to identify value in companies that others might overlook. One famous story is how Munger and Buffett analyzed the financials of See's Candies and concluded that its consistent earnings were more valuable than its current market price suggested. Their ability to crunch the numbers and look beyond the surface was a key factor in Berkshire Hathaway's massive success.

For Product Managers, embracing the role of a "number cruncher" isn't about sitting in front of a pile of spreadsheets with a calculator in hand (although that can happen). It's about interpreting the data to find the story behind it—like figuring out which metrics will truly move the needle for your product, while navigating a sea of "urgent" feature requests (sorry, Chad, the cat mood tracker can wait). Sometimes, it feels like you're piecing together a complex puzzle with just a handful of numbers and a lot of coffee. But when you finally uncover that game-changing insight, it's like discovering the hidden treasure buried within a mountain of data—and suddenly, all the effort feels worth it.

118. WHIZ KID

WHAT/WHEN

A young person with intellectual abilities beyond their age.

> **"Bill McNamara was a whiz kid at Ford Motor Company during WWII. He eventually became the country's Secretary of Defense."**

AS A PRODUCT MANAGER

Being the smartest person in the room isn't always as glamorous as it sounds. This idiom is a badge worn by those who seem to have an innate ability to solve problems before anyone even realizes they exist. A "whiz kid" is often someone who is effortlessly good at what they do, especially in a way that makes the rest of us feel slightly inadequate. In industries like tech, these are the people who can solve complex algorithms in their sleep or design an entire app before lunch—and they don't even need coffee to do it.

One classic example of a "whiz kid" moment is when Mark Zuckerberg was just a college sophomore, building the first version of Facebook in his dorm room. His ability to create an entire social media platform that would change the world—while balancing finals and roommate drama—was nothing short of wizardry. While other students were stressing over term papers, Zuckerberg was busy redefining how humans connect, all from the comfort of his dorm desk.

For Product Managers, being a "whiz kid" means having that magical ability to come up with a brilliant solution out of thin air—like when a team member asks how to fix a process that's been broken for months and you solve it before they finish their sentence. It's a mix of brains, charm, and pure wizardry that makes people think, "Wait, how did you do that?" Of course, you'll have the occasional awkward moment when you don't know the answer to something, but hey, nobody's perfect. Just pull out your mental toolbox, throw out a vague but confident solution, and watch the team's jaws drop. It's less about perfection and more about making everyone think you're the genius who knew it all along.

119. GIFT OF GAB

WHAT/WHEN

Having a facility with words.

> **"Joey could sell snow in Minnesota. He has the gift of gab."**

AS A PRODUCT MANAGER

When it comes to getting your point across, there's nothing like having the "gift of gab." This idiom describes the ability to speak fluently and persuasively, often with a touch of charm that makes even the dullest topics sound interesting. It's a powerful tool in almost any conversation, whether you're pitching a new idea, negotiating a deal, or just trying to convince your friend to split the dessert.

One famous example of someone with the "gift of gab" is none other than former U.S. President Bill Clinton. Known for his smooth, eloquent speeches and ability to connect with people from all walks of life, Clinton's charm was as much a part of his political identity as his policies. His gift for gab helped him navigate tricky situations and build relationships that carried him through both successes and scandals.

For product managers, the "gift of gab" is less about telling a great story and more about talking your way out of a jam. It's about making sure everyone on the team is on the same page even when they're speaking entirely different languages. Imagine having to explain to a developer why the client wants that "one tiny change" in the product, all while keeping a smile on your face and not breaking into a sweat. Sometimes, it feels like you're not just managing a product; you're managing expectations, egos, and the occasional crisis—all without missing a beat. But when you've got the gift of gab, it's all part of the charm.

120. HE WHO HESITATES IS LOST

WHAT/WHEN

Indecision often leads to failure.

> **"When it comes to closing deals swiftly, remember that he who hesitates is lost."**

AS A PRODUCT MANAGER

You've probably heard the phrase "he who hesitates is lost" tossed around in stressful situations. This idiom highlights the importance of decisiveness, especially when time is of the essence. In the fast-paced world of business or even daily life, waiting too long to make a decision can lead to missed opportunities or, worse, a complete collapse of your plans. It's about striking while the iron is hot, even if you're unsure if you're holding the right tool.

A perfect example of "he who hesitates is lost" in action can be seen in the fast-paced world of startups. Take the story of Airbnb, where co-founders Brian Chesky and Joe Gebbia had to make quick decisions when they were struggling to keep the company afloat in its early days. When they realized they had to take immediate action to secure funding or risk running out of money, they didn't wait around for the perfect pitch. Instead, they created a prototype to show potential investors, sealing their first big investment and saving the company from failure.

For Product Managers, hesitation can feel like watching a train speeding toward you while you're still trying to figure out which track you should be on. It's that moment when you're deciding whether to make a change to the app just before launch or to push it out and risk customer backlash. The reality? You don't have time to deliberate. Just like in the startup world, sometimes the best decisions are made in a hurry—because, at the end of the day, the train doesn't stop for anyone.

121. BEAT AROUND THE BUSH

WHAT/WHEN

An inability to get to the point.

> **"Jill's idea makes no sense, but Mike worries about hurting her feelings. He's just beating around the bush."**

AS A PRODUCT MANAGER

Sometimes, we all have a tendency to avoid the tough stuff, opting instead for a long-winded explanation or detour that avoids the heart of the issue. This is where the idiom "beat around the bush" comes in—meaning to avoid addressing the main point directly. Whether it's in a meeting or in everyday conversation, we've all encountered someone who prefers to talk around a problem rather than tackle it head-on.

A good example of this is with the company Pebble, a smartwatch startup that gained popularity back in the early 2010s. Pebble's product was initially well-received, but they struggled to keep up with larger competitors like Apple and Fitbit. Instead of directly addressing their declining sales and the fact that their product couldn't compete with the newer tech on the market, Pebble's leadership often talked about their community support, Kickstarter success, and how they were "re-imagining the wearable experience." They beat around the bush instead of confronting the fact that their product was outdated and losing traction, which ultimately led to their acquisition by Fitbit in 2016.

For Product Managers, beating around the bush is a dangerous game, especially when your product's future is on the line. Imagine being asked if a new feature is ready for launch and instead of saying, "No, it's not," you start rambling about how "there's a lot of exciting potential with this feature." While that might sound positive, it doesn't solve the problem or answer the question. It's like talking about your product's strengths when your customers are already pointing out the weaknesses. Sometimes, it's better to just get straight to the point and admit when things aren't quite where they need to be, so you can get to work fixing them.

122. BETTER LATE THAN NEVER

WHAT/WHEN

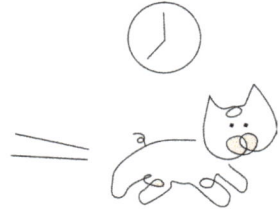

The recognition that performance is better than non-performance.

> **"We're going to miss the kick-off date. I suppose it's better late than never."**

AS A PRODUCT MANAGER

Sometimes, it's not about being the fastest, but about showing up and getting the job done, even if you're a little late. The idiom "better late than never" speaks to the idea that it's more important to act, even if you're behind schedule, than to not act at all. In the fast-paced world of product management, this could be about making sure a feature is released eventually, even if it didn't make the original deadline.

A perfect example of this comes from the world of tech giants like Google, who famously took years to launch Google Drive. The company was late to the cloud storage game, while competitors like Dropbox had already gained a huge following. But in the end, Google's solution wasn't just a copy—it had extra perks that made it a game-changer. While Google was technically late to the party, they made sure that when they arrived, they had something to offer that was better than what everyone else brought.

For Product Managers, "better late than never" often means cutting through the pressure of missed deadlines and the "what-if" scenarios to deliver something valuable—just a little later than expected. It's like showing up to a party with the perfect dish when everyone else is already halfway through dessert. You might be fashionably late, but you'll still get props for bringing the crowd-pleasing main course. In the end, it's not about when you show up; it's about making sure what you deliver is worth the wait.

123. CUT YOUR LOSSES

WHAT/WHEN

Recognition that a product, project, or effort has failed.

> "We've added features, lowered the price, and over-spent the ad budget. Nothing has worked. It's time to cut our losses."

AS A PRODUCT MANAGER

Sometimes in the corporate world, the wisest move isn't about doubling down—it's about knowing when to walk away. This idiom captures the essence of recognizing a failing endeavor and having the courage (and spreadsheet data) to say, "Enough is enough." In product management, this might mean pulling the plug on a flop of a feature before it drains even more time, money, and team morale.

One real-world example of a product manager "cutting losses" is when Peloton faced a drastic shift in demand once the pandemic-era boom subsided. After ramping up production to meet skyrocketing sales, the company suddenly found itself with more bikes than enthusiastic buyers. Rather than keep forcing inventory on a cooling market, they scaled back manufacturing and redirected their strategy—painful, but better than drowning in stationary bikes nobody wanted.

For Product Managers, "cutting your losses" is less about waving a white flag and more about preventing a small leak from sinking the whole ship while you're busy debating the color of the lifeboats. It's about spotting that point of diminishing returns—like bailing on a never-ending redesign that nobody asked for (no, Chad, your third pivot in four sprints isn't going to save it). Sometimes, it feels like you're the captain, the storm, and the confused weatherman all at once. But hey, nothing beats the sweet relief of setting aside a doomed project so you can focus on the next big (and hopefully more seaworthy) idea.

124. BLESSING IN DISGUISE

WHAT/WHEN

A setback that becomes an advantage.

> "We rushed the product to market and overlooked the battery problems. The problems turned out to be a blessing in disguise by avoiding similar problems in the future."

AS A PRODUCT MANAGER

It's easy to assume that a failure is just a failure, especially when things seem to go south fast. But the phrase "blessing in disguise" reminds us that sometimes, the most unexpected twists can lead to success. In the product world, this might happen when a seemingly bad setback actually pushes the product in a better direction than you could have imagined.

Take the story of Angry Birds as a perfect example of a "blessing in disguise." The game started as a flop in its early stages. Initially, the developers at Rovio had trouble making their game work and didn't see the kind of traction they had hoped for. However, a complete overhaul of the game led to the creation of the iconic angry birds and their slingshot-based mechanics. What began as a simple, unremarkable idea turned into one of the most successful mobile games ever.

For Product Managers, finding a "blessing in disguise" can feel like striking gold after hitting a rock bottom. Sometimes a failed feature or a design that no one thought would work can end up being the breakthrough moment for your product. The key is embracing the chaos, because in the world of product management, the road to success isn't always straight—sometimes, it's the detours that take you to the best destination.

125. BITE THE BULLET

WHAT/WHEN

Accepting the consequences of a bad decision.

> **"We never should have tried to compete with an inferior product. We need to bite the bullet, cut our losses, and concentrate on a new product."**

AS A PRODUCT MANAGER

There are moments in business when you know you have to do something hard, and you just have to power through it. That's where the idiom "bite the bullet" comes in. It's about facing up to tough situations and pushing forward, even when the outcome might not be ideal. In the world of small businesses, this could mean dealing with a tough customer complaint or making the hard decision to change a product line that's no longer selling well.

A great example of this comes from the outdoor gear company Patagonia. In 2011, they made the bold decision to pull their popular "Don't Buy This Jacket" ad campaign, which encouraged consumers to think twice before making unnecessary purchases. While it initially shocked many in the industry, the company bit the bullet by sticking to their values of environmental sustainability. Despite concerns that it would hurt sales, Patagonia's commitment to environmental responsibility ultimately built customer loyalty and increased brand trust.

For Product Managers, "biting the bullet" often means making decisions that are uncomfortable but necessary in the long run. It could be choosing to retire a feature that users aren't responding to, or pushing forward with a price hike that you know will cause some grumbling. It's not easy, but it's about sticking to your principles or making a call that's in the best interest of the company in the long-term, even when it might hurt short-term. Sometimes, biting the bullet is the only way to maintain integrity and keep moving forward.

126. BARKING UP THE WRONG TREE

WHAT/WHEN

Pursuing a mistaken or misguided course of action.

> **"The team spent weeks focusing on the wrong target market, ultimately barking up the wrong tree in their product development strategy."**

AS A PRODUCT MANAGER

Sometimes, we find ourselves chasing down the wrong lead, especially when we're convinced we're right. This idiom perfectly captures the idea of misdirecting efforts toward a dead end. In the world of product management, it's easy to get caught up in assumptions and pursue solutions that aren't even close to addressing the real issue, especially when there's a tight deadline or pressure from stakeholders.

A recent example of "barking up the wrong tree" comes from the food tech startup *Zume*, which aimed to revolutionize pizza delivery with an automated system of trucks equipped with pizza-making robots. The company invested heavily in creating a fleet of these smart trucks, hoping to optimize the pizza-making process and cut delivery costs. However, Zume overlooked more fundamental challenges, such as customer demand for fresh, high-quality pizza and operational scaling issues. The result? Zume ended up pivoting away from the pizza-making trucks and ultimately scaling down its business, realizing that their tech-focused solution wasn't aligned with the actual needs of their target market. They were barking up the wrong tree, thinking that automation was the key to success, while their customers were more concerned about taste, quality, and consistency.

For Product Managers, "barking up the wrong tree" isn't just about making a mistake—it's about being so sure you're on the right path that you end up wasting a lot of time and resources while the solution is sitting in plain sight. It's the product version of obsessing over complex technology while ignoring the simple, fundamental issues that matter most to your customers. And sometimes, after all that time and energy, you realize that a good pizza is all about the ingredients and human touch, not robots. But hey, at least Zume learned the hard way that tech can't fix everything.

127. A BIRD IN THE HAND IS WORTH TWO IN A BUSH

WHAT/WHEN

Don't abandon a sure thing for a possibility.

> "Rather than chasing every new trend, our strategy should focus on strengthening existing partnerships, keeping in mind that a bird in the hand is worth two in the bush."

AS A PRODUCT MANAGER

When making decisions, it's easy to get swept up in the excitement of potential growth, but sometimes the key is to focus on what's already in your hands. The phrase "A bird in the hand is worth two in the bush" reminds us that it's better to hold on to what's working rather than risk everything for something uncertain. In business, this means sticking with a solid, proven approach instead of overextending on a gamble.

A perfect example of this from 2023 is the downfall of Better.com. The digital mortgage platform was booming at first, but the company's aggressive push to scale too quickly—without solidifying its core business model—ultimately led to its struggles. Better.com chased ambitious goals, aiming to revolutionize the mortgage industry, but in the process, they lost sight of the need to first stabilize their core product offering. Instead of perfecting what was already working, they risked everything on an uncertain future, and the results were catastrophic.

For Product Managers, the lesson here is clear: Sometimes, the best move is to double down on what's working rather than reaching for something that could be out of your control. Scaling too quickly without a stable foundation can lead to disaster, just as Better.com found out. It's about ensuring that the bird in your hand is well-cared-for before chasing the next big thing.

128. PAINT WITH A BROAD BRUSH

WHAT/WHEN

Ignoring details to present a broad concept.

> **"Paul has a habit of painting with a broad brush, forgetting the details are what matters."**

AS A PRODUCT MANAGER

When you paint with a broad brush, you're looking at the big picture instead of getting lost in the nitty-gritty details. This idiom emphasizes the tendency to generalize or simplify complex issues, often to make a point or save time. It's like trying to summarize an entire novel in one sentence, hoping that everyone will get the message—without worrying about all the little chapters that might change the story.

A great example of this can be seen in the marketing world during product launches. Take, for instance, when tech companies call their products "game-changing" or "revolutionary" in their ads. They might gloss over the minor flaws or limitations in favor of emphasizing the shiny, appealing features. It's all about the broad strokes, where every product looks perfect for everyone—because who has time for the messy stuff when you're selling an idea, right?

For Product Managers, "painting with a broad brush" often means focusing on big-picture strategies that sound impressive without getting into the weeds. It's like showing up to a meeting with a PowerPoint that says "We're innovative!" without diving into user research or testing—because who has time for that when Chad is asking if he can add one last feature? (Spoiler: No, Chad, we can't.) But hey, it's all good until the product hits the market and you realize you might need a bit more than just a bold headline to make it work.

129. TOP DOG

WHAT/WHEN

The dominant player in a group.

> **"Amazon is the top dog in the online retail industry."**

AS A PRODUCT MANAGER

When it comes to running a successful business, being the "top dog" means taking charge and making the tough calls that shape the company's future. This idiom perfectly captures the responsibility of leadership, where every decision counts and the weight of the company's success or failure falls on your shoulders. In the world of the UFC, the top dog isn't just a fighter; it's the leadership that drives the company forward, making strategic moves and keeping the organization at the top of its game.

One notable example of UFC's top dog leadership was when Dana White took over the company in 2001, turning it from a struggling brand into a global powerhouse. White's decision to push for mainstream acceptance by securing broadcast deals, creating partnerships, and promoting high-profile fighters turned the UFC into a multi-billion dollar business. His ability to take risks, make bold moves, and adapt to the changing landscape made him the undisputed top dog in the fight sports industry.

For a company like UFC, being the "top dog" is all about making tough business decisions—like whether to sign a controversial fighter who can bring attention but also risk reputation—or when to expand globally, entering new markets with potential for growth. It's also about knowing when to lean into the spectacle of the sport and when to focus on building the brand behind the scenes. Leading the UFC means juggling promotional stunts, business negotiations, and fan expectations all at once. It's not always glamorous, but someone has to wear the crown—and keep it on straight while the world watches.

130. STICK OUT LIKE A SORE THUMB

WHAT/WHEN

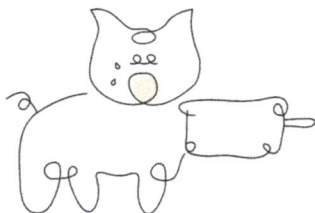

An obvious fault or mistake.

> "The company's ineptness sticks out like a sore thumb. Half of their products don't work."

AS A PRODUCT MANAGER

Blending in is often important in the corporate world, where teams are expected to work seamlessly together. However, there are times when something or someone just can't help but "stick out like a sore thumb." This idiom perfectly describes those instances when something stands out in such a way that it can't be ignored, often for the wrong reasons. In the business world, this could be a flashy new idea that's just too bold for the company culture or a marketing campaign that goes off the rails.

A prime example of a company "sticking out like a sore thumb" is WeWork. In the early days, WeWork's mission to "elevate the world's consciousness" and its over-the-top, almost cult-like culture made it stand out in the co-working space industry. With its lavish office spaces, extravagant parties, and founder Adam Neumann's eccentricities, WeWork quickly became the outlier, turning what was supposed to be a simple office-sharing model into a public spectacle. It wasn't just a workspace; it was a lifestyle, and not everyone was buying into it.

For companies, "sticking out like a sore thumb" is more about being the oddball in the room, where your big, bold moves make everyone else a little uncomfortable. It's like showing up to a board meeting in a full superhero costume—everyone notices you, but maybe not in the way you hoped. Sometimes, it feels like you're the one throwing confetti at a funeral, trying to be the life of the party while everyone else is just trying to get through their spreadsheets. But hey, nothing grabs attention like a sore thumb, even if it's for all the wrong reasons.

131. COPYCAT

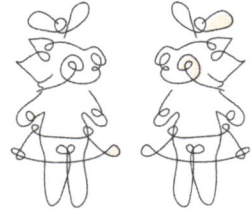

WHAT/WHEN

A clone or near replica of another product.

> **"Why reinvent the wheel when we can copycat it?"**

AS A PRODUCT MANAGER

Imitating others might seem harmless at first, but it's often a sign of laziness or lack of originality. The phrase 'copycat' highlights this tendency to simply mimic someone else's work instead of creating something new. In the world of business, this idiom can describe a company that takes inspiration from a competitor's product without adding any innovation or unique value.

A prime example of a 'copycat' in action comes from the tech world, where companies frequently replicate successful features. Take the case of Facebook's launch of its Stories feature in 2017, which was clearly modeled after Snapchat's popular Stories. Facebook didn't reinvent the wheel but instead borrowed the idea to capitalize on its already massive user base. While they eventually became a success, it wasn't exactly a moment of groundbreaking originality.

For Product Managers, dealing with 'copycats' is like being stuck in a brainstorming session where someone keeps suggesting the same ideas over and over but with no new twist. It's the frustrating reality of facing competitors who watch your every move and try to replicate it, as if a product's value lies in just copying the design. Sometimes, it feels like you're Chad, having spent months developing something unique, and then someone else swoops in with a nearly identical idea, asking, 'Hey, can we add one more feature to this?' (Spoiler: no, Chad, we can't)."

Imitating others might seem harmless at first, but it's often a sign of laziness or lack of originality. The phrase 'copycat' highlights this tendency to simply mimic someone else's work instead of creating something new. In the world of business, this idiom can describe a company that takes inspiration from a competitor's product without adding any innovation or unique value.

A prime example of a 'copycat' in action comes from the tech world, where companies frequently replicate successful features. Take the case of Facebook's launch of its Stories feature in 2017, which was clearly modeled after Snapchat's popular Stories. Facebook didn't reinvent the wheel but instead borrowed the idea to capitalize on its already massive user base. While they eventually became a success, it wasn't exactly a moment of groundbreaking originality.

For Product Managers, dealing with 'copycats' is like being stuck in a brainstorming session where someone keeps suggesting the same ideas over and over but with no new twist. It's the frustrating reality of facing competitors who watch your every move and try to replicate it, as if a product's value lies in just copying the design. Sometimes, it feels like you're Chad, having spent months developing something unique, and then someone else swoops in with a nearly identical idea, asking, 'Hey, can we add one more feature to this?' (Spoiler: no, Chad, we can't)."

132. EGG ON YOUR FACE

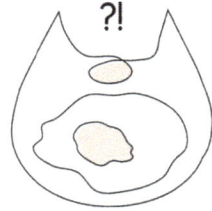

WHAT/WHEN

Feeling embarrassed or looking foolish due to a mistake or blunder.

> **"I promised that the ad campaign would be ready for the product launch. The delay has left me with egg on my face."**

AS A PRODUCT MANAGER

Sometimes, even the most promising innovations can end up leaving you with egg on your face. This idiom is perfect for describing those cringe-worthy moments when something you thought would work out brilliantly ends up being a bit of a flop. In the world of tech, it's especially true when you're working on cutting-edge products that promise to change the game—only to find out later that the hype doesn't quite match the reality.

Take Magic Leap, for instance. When they unveiled their augmented reality headset, the world was buzzing with excitement, anticipating a revolutionary new way of interacting with the digital world. The hype was massive—investors were eager, and the media was enamored. But when the product finally launched, it didn't live up to the grand promises. Users were underwhelmed by the performance, and it quickly became clear that the magic everyone was expecting had failed to materialize. Magic Leap had egg all over its face, and the road ahead was filled with tough lessons.

For Product Managers, "egg on your face" is something you hope to avoid, but sometimes, it's just part of the journey. It's like putting all your energy into something that seems perfect on paper, only for it to come crashing down in front of you. With Magic Leap, the company had to quickly pivot and adjust its strategy after the initial disappointment. The lesson? In tech, even the most ambitious projects can stumble, and sometimes all you can do is wipe the egg off your face and figure out the next move.

133. LEAN IN

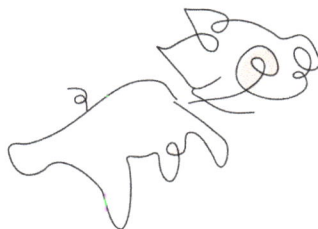

WHAT/WHEN

Encouragement without endorsement to keep options open.

> **"Joe's idea seems to have merit, but I have some questions. I'm going to lean in, but not too far."**

AS A PRODUCT MANAGER

When someone tells you to "lean in," they're not asking you to physically get closer to a conversation (unless you're at a karaoke bar and your friend has had one too many). This idiom is all about embracing challenges with enthusiasm and commitment, especially when they push you out of your comfort zone. In the business world, it's about stepping up, taking charge, and pushing through discomfort to achieve success, whether you're taking on a huge project or tackling a difficult negotiation.

A great example of someone "leaning in" happened in 2014 when Sheryl Sandberg, COO of Facebook, published her book *Lean In*. The book's message encouraged women to push against societal norms and take bold strides in their careers. This became the rallying cry for those looking to break barriers in male-dominated spaces, reminding us that leaning in isn't just about working harder; it's about asserting yourself and demanding a seat at the table—even if the table's in the corner and you're worried about knocking over someone's latte.

For Product Managers, "leaning in" is about making sure you're not the one sitting on the sidelines watching everyone else dive into the deep end. It's about taking the lead when the product is veering toward disaster and being the one who says, "I'll figure this out"—even if you secretly don't know how yet. It's like jumping into a meeting where everyone's talking about their "actionable insights" and pretending to understand, all while frantically googling what the hell a "pivot" is. But you keep leaning in, because that's how you make magic happen, even if the magic feels more like juggling flaming chainsaws than anything remotely glamorous.

134. BEST OF BOTH WORLDS

WHAT/WHEN

An ideal solution to a dilemma or conflict.

> "We've got the best of both worlds. We're the highest quality in the industry with the lowest prices."

AS A PRODUCT MANAGER

It's always satisfying to find a solution that gives you the best of both worlds, especially when you can enjoy the perks of two things that are often hard to combine. This idiom highlights that perfect balance where you get everything you need without sacrificing anything important. In the world of freelancing, it can mean getting access to high-quality services without the steep prices that usually come with professional agencies.

Take Fiverr, for example. Fiverr offers a marketplace where businesses and individuals can find freelance professionals to handle tasks like graphic design, writing, or video editing, all at affordable prices. You get the expertise and skill of a seasoned professional but without the hefty fees you'd pay an agency. Whether you need a logo designed, a product description written, or a catchy social media ad, you can find it on Fiverr—it's like hiring an expert without breaking the bank.

For Product Managers at Fiverr, achieving the "best of both worlds" means balancing the needs of freelancers with the expectations of clients. Imagine you're rolling out a new feature that allows clients to filter freelancers by reviews, skill level, and pricing. You want it to be simple and efficient, but also ensure that freelancers like Chad (who insists on adding a "cool animation" to every single gig, even when it's not needed) can still showcase their talents without confusing the clients. It's all about finding the sweet spot: a seamless user experience for clients and a platform that lets freelancers shine—without getting too carried away with those unnecessary animations.

135. DON'T JUDGE A BOOK BY ITS COVER

WHAT/WHEN

Appearances can be deceiving.

> "Charlie seems to be a nice guy, but he will cut you off at the knees to get ahead. You can never judge a book by its cover."

AS A PRODUCT MANAGER

It's easy to form opinions based on first impressions, but the saying "don't judge a book by its cover" is a reminder that there's often much more beneath the surface. This idiom speaks to the fact that initial perceptions can be deceiving. In the business world, what may seem like a risky or odd choice on the outside can actually turn out to be a brilliant decision when you dig deeper.

Take David Solomon, CEO of Goldman Sachs, as an example. When he was appointed CEO in 2018, many were skeptical. Solomon was known for his background in investment banking, but also for his side hustle as a DJ, which led some to question his seriousness about leading a major financial institution. However, over time, Solomon's leadership style, which balanced traditional finance with an openness to innovation, proved successful. His ability to connect with people from all walks of life and think creatively allowed him to navigate the challenges of leading one of the world's most influential investment banks.

For Product Managers, "don't judge a book by its cover" is a reminder that what seems unconventional or out-of-place at first might hold untapped potential. Sometimes, it's the unexpected combination of skills, experiences, or strategies that leads to success. Just like Solomon's unique path to Goldman Sachs' leadership, your product or team might not always conform to the usual norms. But with the right leadership and vision, even the most unlikely "covers" can lead to incredible outcomes.

136. GETTING OVER YOUR TIPS (OF SNOW SKIS)

WHAT/WHEN

The state of being out of control, typically from moving too quickly.

> "We got over our tips on this one. We should have had another test when the data was inconclusive."

AS A PRODUCT MANAGER

New products are always exciting, loaded with new capabilities and features sure to be a market hit. Optimism and enthusiasm are traits of most PMs and are generally necessary to be successful. Occasionally, enthusiasm is a fault, discouraging realistic evaluation and contrary opinion. PMs need a countering voice on their teams, someone that adds a note of caution about expectations.

In 2020, Volkswagen introduced the Phaeton, a luxury model intended to open a new market segment for the German car manufacturer. By all accounts, the model was a technical success with innovations in engineering, styling, and comfort. Nevertheless, the new model failed to reach its sales target or transform the company's image—moderately priced, well-engineered, small cars—because management failed to understand its brand and market niche.

137. SILVER LINING

WHAT/WHEN

A positive benefit of a disappointing result.

> "Despite the initial setbacks, the silver lining is that our new marketing strategy has significantly increased customer engagement."

AS A PRODUCT MANAGER

It's easy to feel like everything's falling apart when things don't go according to plan, but sometimes the toughest challenges come with hidden opportunities. This idiom perfectly captures the notion that even in the midst of adversity, there's always a silver lining waiting to be discovered. In the SaaS world, this could mean turning a product's early failure into valuable insights or finding a way to capitalize on unexpected market shifts that could redefine your business strategy.

A great example of a "silver lining" in the SaaS world comes from Dropbox. When the company first launched, it struggled to find a paying customer base—most people were hesitant to switch to cloud storage. But during the 2008 financial crisis, businesses started to look for more cost-effective ways to manage data, and Dropbox was positioned perfectly to meet this need. The company capitalized on this shift by offering a free tier to attract users, which eventually helped them build a massive, loyal customer base. What seemed like a slow start ended up being the perfect storm for growth.

For Product Managers in a SaaS company, finding the "silver lining" is often about identifying potential in the middle of setbacks. It's like when a feature you spent months developing doesn't get the traction you expected, but through customer feedback, you discover a completely different use case that no one had considered before. Those moments of pivoting aren't always easy, but they often end up being the moments that redefine your product's future. You're not just surviving the tough times—you're hunting for the hidden opportunities that will turn your obstacles into the next big success.

138. OTHER FISH IN THE SEA

WHAT/WHEN

Unlimited options and opportunities despite failure.

"We failed to get the contract with Walmart.
Thankfully, there are other fish in the sea. Let's
concentrate on the Amazon meeting coming up."

AS A PRODUCT MANAGER

It's important to remember that not every opportunity is your one shot at success, especially when things don't work out as planned. This idiom perfectly captures the idea that there are always alternative routes or options, even if your first choice slips through your fingers. In the tech world, it's tempting to focus on a single feature or a specific customer segment, but "there are other fish in the sea" reminds us that there's a whole market full of potential users, partnerships, and creative pivots waiting to be explored.

One real-world example of a product manager discovering "other fish in the sea" occurred at a software company that provides monthly tools for businesses. They spent months developing a flashy reporting feature, convinced it would be a major selling point. However, initial user feedback made it clear that most customers just wanted a simpler, cleaner interface—no fancy bells or whistles needed. Rather than scrapping everything, the team shifted gears: they took the core ideas from the flashy report and streamlined them into an intuitive dashboard that ended up being a user favorite. By recognizing there was another way to hook their audience, they reeled in new customers and kept existing ones satisfied.

For Product Managers, embracing "other fish in the sea" is less about giving up and more about staying open-minded when your dream feature goes belly up right before launch. It's that moment when you realize you might have to shelf the all-in-one AI chatbot your developers spent weeks building, because your customers really just want a straightforward way to track their daily tasks without the sparkle and glitter. Sure, it can sting to let go of a vision, but once you cast a wider net, you'll often catch something better than you originally expected. It's all about finding fresh angles, pivoting without panicking, and remembering that if one door closes, there's usually an entire ocean of opportunity waiting.

139. BITE OFF MORE THAN YOU CAN CHEW

WHAT/WHEN

Making commitments that cannot be met.

> **"You bit off more than you can chew. Why did you agree to the deadline when you knew half of the team members were on vacation?"**

AS A PRODUCT MANAGER

It's easy to get swept up in the excitement of a big idea, especially when it promises to revolutionize an entire industry. This idiom perfectly captures the moment when enthusiasm leads to overconfidence, resulting in taking on more than you can handle. In the tech and entertainment sectors, it could mean diving into a massive project without fully understanding the scope or challenges involved.

A prime example is Quibi, the short-form streaming service that launched in 2020 with a $1.75 billion investment. The platform aimed to change the way people consumed content, offering high-quality videos designed specifically for mobile devices. But despite the hype and a star-studded lineup, Quibi's lofty ambitions fell short, as the app struggled to find an audience and ultimately shut down in just six months.

For Product Managers, "biting off more than you can chew" is like launching a product with a mountain of expectations but realizing halfway through that you didn't even have the right shovel to get started. It's about thinking you can change an entire market with a flashy new idea, only to realize you didn't anticipate the need for a solid user base, proper timing, or just basic functionality (and no, Chad, we can't fix the UI overnight). It's like trying to build a skyscraper on a foundation of wet sand—ambitious but ultimately doomed unless you scale back and regroup.

140. BREAK THE ICE

WHAT/WHEN

The initial start of a project or conversation.

> **"At the start of the meeting, the team leader shared a light story to break the ice and ease everyone into the discussion. seem to be stuck without a decision."**

AS A PRODUCT MANAGER

Starting a conversation can be tricky, especially when you're meeting someone for the first time. It's crucial to break the ice in order to make things less awkward and set the tone for a comfortable exchange. This idiom highlights the need to ease tension and encourage communication, whether at a high-stakes business negotiation, a first date, or a casual coffee meetup.

One real-world example of "breaking the ice" in the business world is from 2014, when Satya Nadella became the CEO of Microsoft. To ease the tension surrounding the leadership transition, Nadella introduced an "Ask Me Anything" session, allowing employees to openly voice questions and concerns. This small yet powerful gesture helped create an environment where people felt comfortable speaking up, ultimately setting the stage for more transparent and collaborative communication at Microsoft.

For Product Managers, "breaking the ice" might be more about easing friction during stakeholder updates or cross-functional stand-ups than orchestrating grand comedic acts. Sometimes, it involves dropping a well-timed pun about the backlog or acknowledging the universally perplexing nature of "business-casual" dress codes—anything to get everyone feeling less like they're tiptoeing on broken glass. It's a subtle yet vital skill that can transform a stuffy status meeting into a warm, collaborative exchange, because when the air is open and friendly, good ideas have the space to surface without awkwardness.

141. DRIVE A HARD BARGAIN

WHAT/WHEN

The unwillingness to compromise, sometimes to the detriment of a project.

> **"I hate working with those consultants. They drive a hard bargain on their fees, unwilling to give up an inch."**

AS A PRODUCT MANAGER

Negotiating a good deal isn't just about agreeing on a price—it's about knowing how to get the most bang for your buck. This idiom captures the art of striking a deal where everyone thinks they've won, but you walk away with just a little extra. In business, driving a bargain can mean negotiating contracts, making trade-offs, or even getting better perks than initially offered.

A perfect example of "driving a bargain" happened when Mark Cuban bought the Dallas Mavericks. Cuban wasn't just looking to own a sports team—he was looking to turn a losing franchise into a profit machine. He negotiated not just the price of the team, but also secured better terms for the team's operations and fan engagement strategies. Today, the Mavericks are a top contender both on the court and in their business dealings.

For Product Managers, "driving a bargain" is about getting the best deal on everything, from vendor contracts to internal resources. It's less about haggling at a market stall and more about navigating complex negotiations where the stakes are high and the terms are often hidden in fine print. Like getting a third-party vendor to lower their fees without sacrificing quality, while trying to avoid your team's "Hey, can we just throw in an extra feature?" requests. And of course, there's always Chad, who thinks you can "just add one more feature" without affecting the budget or timeline. Spoiler: No, Chad, you can't.

142. BANG FOR THE BUCK

WHAT/WHEN

A measure of impact versus expenditure.

> **"Our decision to offer a twelve-month warranty instead of the industry's six months gave us a lot of bang for the buck."**

AS A PRODUCT MANAGER

When you're looking to make the most out of every dollar spent, the phrase "bang for the buck" is the one that sums it all up. This idiom highlights the need for getting maximum value from an investment, especially when resources are tight. In business, it's all about squeezing as much benefit as you can from every penny, whether you're negotiating a supplier deal or making the most of a marketing budget.

A perfect example of "bang for the buck" comes from the fast food industry, where McDonald's found an innovative way to deliver a big meal at a small price. The Dollar Menu became iconic, offering customers the best deal for their money without sacrificing the size of their meal. It was a brilliant move, giving people a reason to come back time and again for a low-cost, high-satisfaction experience.

For Product Managers, focusing on "bang for the buck" is about ensuring that every resource—be it time, money, or effort—delivers a solid return. It's a bit like when you're deciding between a third round of feature requests or just fixing the two major bugs that will actually make the product usable. Sometimes, it feels like you're managing a delicate balancing act—like choosing between the gold-plated stapler and the coffee maker that keeps the team sane. But in the end, it's all about getting the best deal and still managing to keep everyone happy (or at least caffeinated).

143. PICK SOMEONE'S BRAIN

WHAT/WHEN

Asking for advice or information.

> "I have no idea why people avoid commuter flights. I understand you're the travel expert. Can I pick your brain for a minute?"

AS A PRODUCT MANAGER

When it comes to gathering information, there's no better way than to "pick someone's brain." This idiom emphasizes the value of extracting knowledge or ideas from someone who has experience or expertise in a particular area. Whether you're working on a project or trying to solve a tricky problem, asking the right person can make all the difference.

A great example of "picking someone's brain" is when a startup founder meets with an experienced entrepreneur to gain insights on scaling their business. The founder might ask about challenges, strategies, and tips on navigating growth. The experienced entrepreneur could offer valuable advice that saves the startup from common pitfalls and helps them succeed faster. It's all about tapping into someone else's knowledge to avoid reinventing the wheel.

For Product Managers, "picking someone's brain" is like trying to figure out how a seasoned chef would whip up a five-star meal from leftovers. It's about leveraging the wisdom of those who have been there, done that, and probably have the scars to prove it. It's not so much about asking for a step-by-step recipe but rather picking out the nuggets of wisdom that could turn a mediocre product strategy into a winning one. Plus, you'll often leave the conversation feeling like you've just uncovered a treasure chest of tips—while maybe also feeling a little guilty for not having come up with the idea yourself. But hey, that's what collaboration is all about, right?

144. PLAY IT BY EAR

WHAT/WHEN

Proceeding without a plan, reacting to the circumstances.

> **"We didn't consider that customers might want a repeat function. We'll have to play it by ear until we get solid proof."**

AS A PRODUCT MANAGER

You know that feeling when you have a vague idea of what you need to do, but you're not exactly sure how it's all going to unfold? That's where "play it by ear" comes in. It's the art of improvising as you go, adjusting to whatever comes your way. In a fast-moving business environment, it's often about reacting to unexpected opportunities or obstacles without a clear step-by-step plan in place.

A great real-world example of this is how Airbnb got its start. The founders didn't initially have a fully fleshed-out business plan. They just needed to make some quick cash and figured, why not rent out their own apartment as a place for conference-goers to crash? As the company grew, they had to continuously adapt and adjust, figuring things out on the fly as they expanded into new markets and faced unforeseen challenges. It was a classic "play it by ear" scenario that ultimately paid off in a big way.

For Product Managers, "playing it by ear" often means running with an idea even when the details are still fuzzy, and adjusting as you learn more. It's like pushing out a beta version of a product that's not quite polished, then scrambling to fix bugs as users start giving feedback. You know you've got to keep the ball rolling, but you're not entirely sure where it's headed—so you just keep adjusting your grip and hope you don't drop it. Sometimes, it feels like you're building a plane while it's already in the air—except, of course, Chad insists on adding "one more feature" at the last minute, even though the plane is already halfway to its destination. But hey, that's the thrill of flying by the seat of your pants, right?

145. WORD OF MOUTH

WHAT/WHEN

Transmission of oral information.

> **"Few recommendations are as powerful as word of mouth."**

AS A PRODUCT MANAGER

You know that feeling when you have a vague idea of what you need to do, but you're not exactly sure how it's all going to unfold? That's where "play it by ear" comes in. It's the art of improvising as you go, adjusting to whatever comes your way. In a fast-moving business environment, it's often about reacting to unexpected opportunities or obstacles without a clear step-by-step plan in place.

A great real-world example of this is how Airbnb got its start. The founders didn't initially have a fully fleshed-out business plan. They just needed to make some quick cash and figured, why not rent out their own apartment as a place for conference-goers to crash? As the company grew, they had to continuously adapt and adjust, figuring things out on the fly as they expanded into new markets and faced unforeseen challenges. It was a classic "play it by ear" scenario that ultimately paid off in a big way.

For Product Managers, "playing it by ear" often means running with an idea even when the details are still fuzzy, and adjusting as you learn more. It's like pushing out a beta version of a product that's not quite polished, then scrambling to fix bugs as users start giving feedback. You know you've got to keep the ball rolling, but you're not entirely sure where it's headed—so you just keep adjusting your grip and hope you don't drop it. Sometimes, it feels like you're building a plane while it's already in the air—except, of course, Chad insists on adding "one more feature" at the last minute, even though the plane is already halfway to its destination. But hey, that's the thrill of flying by the seat of your pants, right?

146. BELT AND SUSPENDERS MAN

WHAT/WHEN

Description of an extremely cautious person.

> **"He's a belt and suspenders man, always implementing extra safeguards to ensure the project's success."**

AS A PRODUCT MANAGER

In the fast-paced world of product design, some people take a methodical approach to avoid any missteps—double-checking, triple-checking, and even running through plans one more time, just in case. The "Belt and Suspenders Man" idiom reflects the cautious individual who makes sure every base is covered, even if it seems a bit excessive. In the design industry, this often means making sure everything from wireframes to prototypes has multiple backups to ensure things never go wrong.

A prime example of the "Belt and Suspenders Man" mentality can be seen during Figma's rollout of its real-time collaboration features. The design team had to ensure that not only was the design interface smooth and intuitive, but that multiple layers of redundancy were in place to prevent any hiccups with real-time editing. They set up fallback systems to ensure users could still collaborate without interruptions, even if one part of the service had a hiccup. While some might have found this overkill, it guaranteed a seamless experience for millions of users.

For Product Managers, being a "Belt and Suspenders Man" means obsessively preparing for every worst-case scenario. It's like having a backup plan for your backup plan, ensuring no loose ends. Imagine working on a design feature for weeks, only to realize you've accounted for every tiny glitch in the system—just in case someone clicks the wrong button. You might not get the recognition for being the safety net, but you're definitely the unsung hero who prevented the whole thing from crashing down when a small bug decided to throw a tantrum.

147. LET THE CAT OUT OF THE BAG

WHAT/WHEN

Failure to keep a secret.

> "Forget about the layoff surprise. Mary has already let the cat out of the bag."

AS A PRODUCT MANAGER

It's easy to get excited about new ideas, but there's something to be said for keeping secrets, too. This idiom highlights the importance of discretion and timing, reminding us that some things are best kept under wraps until the right moment. In the world of product management, this could apply to everything from a surprise product launch to the inevitable "oops, we accidentally released a bug" moments.

Take the case of the 2013 launch of the iPhone 5S, which had a little surprise waiting for consumers—Touch ID. Apple managed to keep this feature a secret until the big reveal, building up suspense and anticipation. If anyone had "let the cat out of the bag" early, it would've ruined the magic and excitement of the announcement. A misstep like that could have diminished the impact of one of Apple's most game-changing features.

For Product Managers, "letting the cat out of the bag" is like revealing the punchline to a joke before you've even told the setup. Sometimes, it's about knowing when to zip it—whether you're keeping hush-hush about a new feature or dealing with the temptation to share something prematurely. And of course, there's always Chad, who's *just* dying to tell everyone about the shiny new feature he's been working on for months (yes, Chad, we know you're excited—but we'll *let* you share it when the time is right). Keeping things under wraps isn't just about playing coy; it's about knowing that once that cat's out of the bag, there's no putting it back in. But hey, nothing gets a team quite as motivated as the thrill of a perfectly timed reveal—just make sure Chad doesn't spill the beans before the showtime!

148. CHICKEN OR EGG

WHAT/WHEN

Confusion over priorities.

> "The debate over whether to invest in marketing or product development first is a classic chicken or egg scenario in business strategy."

AS A PRODUCT MANAGER

When trying to decide the best approach to growth, it's important to identify which aspect needs the most attention. The "chicken or egg" idiom captures that feeling of being caught in a cycle of dependency, where it's unclear whether to focus on building the product or growing the user base first. This is especially true for social media platforms, where both are essential to success but rely heavily on each other. In LinkedIn's case, the challenge was deciding whether to improve features for users or focus on attracting more professionals to the platform.

A great example of LinkedIn facing the "chicken or egg" dilemma was when the platform was first launched. It had the challenge of convincing professionals to sign up and create profiles, but without a critical mass of users, there was little incentive for anyone to join. Should LinkedIn focus on creating more user-friendly features to attract signups, or should they invest in reaching out to more professionals to make the network valuable?

Eventually, LinkedIn realized it had to walk the line between both, improving features while simultaneously pushing hard to expand its user base.

For product managers at LinkedIn, the "chicken or egg" situation felt like a constant balancing act. Do you spend your resources building more robust networking tools for users, or do you pour your energy into bringing in new professionals to make those tools useful? In the early days, it was a bit like getting stuck in a loop of "build it, and they will come" vs. "if they come, we'll build it." LinkedIn's breakthrough came when they realized they needed both to grow simultaneously, understanding that having the right features was meaningless if no one was around to use them—and vice versa. It's a classic case of juggling two equally important priorities until they finally sync up.

149. KEEP AN EAR TO THE GROUND

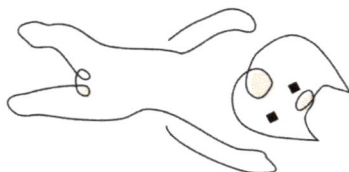

WHAT/WHEN

Being actively aware of information that impacts the objective.

> **"We need to keep an ear to the ground for any emerging market trends that could impact our product strategy."**

AS A PRODUCT MANAGER

It's essential to stay in the loop, particularly when you're trying to stay ahead of the curve. "Keep an ear to the ground" is a perfect idiom to capture the essence of being tuned into what's happening around you. In the fast-paced world of business, it's all about catching early whispers and anticipating what's coming next, whether it's a market shift or the latest trend in customer demands.

A great example of someone "keeping an ear to the ground" would be the legendary Jeff Bezos during the early days of Amazon. As the company was growing, Bezos was known for his keen awareness of customer feedback and the shifting needs of the e-commerce landscape. He was always listening—literally and figuratively—whether it was to customer complaints or market signals, so that Amazon could adapt quickly and stay ahead of its competition.

For product managers, "keeping an ear to the ground" isn't about being nosy, but more about hearing the faint buzz of potential opportunities or pitfalls before they explode into full-blown issues. It's the skill of knowing that if the office coffee machine is suddenly silent, something's up. Maybe it's a new feature request from an important client, or perhaps it's just Chad forgetting to refill the water cooler again (seriously, Chad). Either way, being alert to those subtle cues lets you dodge the big problems before they hit you like a freight train, all while keeping your sanity intact.

150. NEEDLE IN A HAYSTACK

WHAT/WHEN

An outcome or information that appears impossible to achieve or find.

> **"Finding bad code in 1000 lines of programming is like finding a needle in a haystack."**

AS A PRODUCT MANAGER

Finding something as specific as a "needle in a haystack" can feel like an impossible task, especially when you're dealing with a massive security breach. This idiom highlights the challenge of tracking down that one rare, critical flaw hidden among a sea of other possibilities. In the world of cybersecurity, it often means identifying the single point of vulnerability that allowed a malicious actor to infiltrate a system, causing a cascade of problems.

A real-world example of this is the 2017 Equifax data breach, where hackers exploited a known vulnerability in the company's system. Despite the patch being available, Equifax failed to apply it in time, and the attackers were able to access the personal data of millions of people. Security teams had to conduct a thorough search through massive amounts of data and logs, like trying to find a needle in a haystack, to trace the breach back to its origin and understand its full impact.

For product managers and cybersecurity teams, hunting for a "needle in a haystack" during a cyberattack means juggling the chaos of a breach while trying to fix things under a crushing time crunch. It's a process of sifting through tons of irrelevant information to pinpoint exactly where things went wrong, all while dealing with mounting pressure and a team of people demanding answers. It's not glamorous, and it's rarely a clean win. But when the needle is found, it's a moment of clarity amid the mess—and a reminder that even the smallest vulnerability can cause a huge problem.

151. BET THE FARM

WHAT/WHEN

A desperate effort in the hope of success.

> **"The CEO decided to bet the farm on the new product launch, investing the company's entire budget to ensure its success."**

AS A PRODUCT MANAGER

When you're fully committed to a risky venture, you might find yourself "betting the farm." This idiom expresses putting everything on the line, whether it's resources, time, or energy, in hopes that it pays off. In business, this might mean making a huge investment in a new product, strategy, or partnership with the potential for a massive reward—or a disastrous loss.

A prime example of someone "betting the farm" is Elon Musk with the early days of SpaceX. In 2008, Musk was down to his last $30 million after three failed launches. Rather than cut his losses, he took a massive risk, betting everything on a fourth launch, which eventually succeeded and secured NASA contracts, turning SpaceX into the powerhouse it is today. Had that launch failed, SpaceX would have been out of business, and Musk's reputation might have crumbled.

For Product Managers, betting the farm is all about trusting your gut even when the odds are stacked against you. It's like deciding to roll with an untested feature that could either make your product soar or crash and burn—while the whole team is watching. It's not about the big wins you share in a victory lap, but about those heart-stopping moments when you're one bad decision away from a full-on disaster. You just hope you can cash in before you realize your "bet" was a bit too risky for comfort.

152. SPANNER IN THE WORKS

WHAT/WHEN

An unexpected obstacle to success.

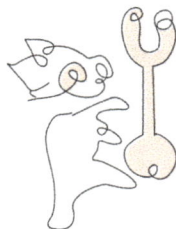

> **"The unexpected change in regulations threw a spanner in the works, delaying our product launch by several weeks."**

AS A PRODUCT MANAGER

When everything is running like clockwork, it can feel like you're on autopilot. But then, out of nowhere, someone throws a spanner in the works, and suddenly, things are anything but smooth. This idiom captures the frustrating but inevitable moment when an unexpected problem messes up even the best-laid plans. In the tech world, it could mean a sudden product recall, an unexpected hardware malfunction, or even a competitor's surprise product launch that forces a change in strategy.

A great example of this happened with NVIDIA during the launch of their RTX 30 series graphics cards in 2020. Despite all the hype and excitement, the launch faced major supply chain issues, leading to shortages and scalpers snapping up cards, selling them at massively inflated prices. NVIDIA's product management team had to quickly react to the chaos, working overtime to fix the supply issues and manage frustrated customers. The spanner in the works? Well, it was a combination of high demand, manufacturing delays, and, let's be honest, a bit of bad timing.

For Product Managers, a spanner in the works often means juggling the fallout from multiple angles—product delays, customer complaints, and a PR crisis all in one go. It's like preparing for a smooth launch only to find that half your servers are down, your marketing team is working off outdated specs, and someone on the team—looking at you, Chad—is asking if it's too late to add a "cool new feature" (spoiler: no, Chad, it's definitely too late). You end up being part firefighter, part therapist, and part magician—trying to fix the situation while convincing everyone that yes, it will get better. But hey, at least you'll come out the other side with a great story about how you survived a true spanner-in-the-works moment.

153. WALK A TIGHTROPE

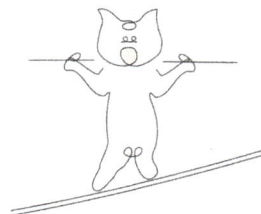

WHAT/WHEN

Deciding between opposing forces.

> **"In managing the AI chatbot project timeline, we had to walk a tightrope between meeting client expectations and staying within budget."**

AS A PRODUCT MANAGER

It's essential to keep things balanced when you're navigating through a situation with high stakes. The idiom "walk a tightrope" captures the essence of managing risk while trying to achieve something great. For companies like Intuit, this often means balancing the development of new features with maintaining the reliability and simplicity of their core products, such as TurboTax or QuickBooks, without overwhelming their user base.

A great example of Intuit walking a tightrope occurred when they transitioned TurboTax to an online platform. The company had to innovate by offering new features and services in the cloud, while still maintaining the ease of use that their customers loved. The risk was high: too many changes could alienate loyal users, but failing to innovate would leave them behind in an increasingly digital world. The solution was a careful, strategic balance of updating the software without sacrificing its user-friendly appeal.

For Product Managers, "walking a tightrope" is about juggling innovation, customer satisfaction, and business goals without letting any one of them topple the others. It's like working on a major product update while still addressing urgent customer support tickets that remind you how much users love the simplicity (and how much they fear change). Balancing constant feedback, internal expectations, and market trends can feel like one wrong move could lead to a crash, but when done right, it results in a product that's both innovative and beloved.

154. SPREAD YOURSELF TOO THIN

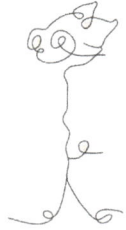

WHAT/WHEN

Accepting too many responsibilities or obligations.

> **"I love Judy's attitude, that get-up-and-go spirit. But I worry she is spreading herself too thin and will end up missing deadlines."**

AS A PRODUCT MANAGER

It's easy to get excited about tackling multiple tasks at once, especially when everything feels important. The idiom "spread yourself too thin" is a fitting way to describe the consequences of overcommitting to too many things. It's that feeling when you're juggling so many projects that you forget which one is about to set off the fireworks display of disaster.

A great example of this is when a product manager at a fast-growing tech startup tries to oversee the launch of three new features in one quarter, while simultaneously managing customer support, organizing an office move, and approving the company's holiday party plans. All of a sudden, things get messy, and you're half-finished with everything—sort of like ordering three different meals at a restaurant but only getting halfway through the appetizers before realizing you're too full to continue. In the end, no one's really satisfied, and you're left with a cold plate of regret.

For Product Managers, "spreading yourself too thin" is all about pretending you can handle everything at once, even though your calendar is already a chaotic puzzle of back-to-back meetings and vague to-do lists. It's like trying to write a product roadmap while answering emails, fixing bugs, and coming up with witty responses for the CEO's LinkedIn post all at the same time. You end up feeling like a circus performer, only without the applause and with much more caffeine. In the end, it's all about prioritizing—because there's no way you can finish that spreadsheet, respond to customer complaints, and research your next vacation destination without at least one of those things getting a little burnt out.

155. CUT OFF YOUR NOSE TO SPITE YOUR FACE

WHAT/WHEN

Acting foolishly due to pride.

> "Firing a key employee out of spite could be a classic example of cutting off your nose to spite your face, as it might hurt the company's productivity and morale in the long run."

AS A PRODUCT MANAGER

Sometimes, people make decisions that seem, well, a little counterproductive. That's where the phrase "cut off your nose to spite your face" comes in—describing the kind of action that's both self-destructive and utterly baffling. It's like deciding to walk out of a meeting early to prove a point, only to realize you missed the only key decision that could've saved your project. In the business world, it's the kind of decision that makes everyone else scratch their heads and wonder, "Did that really just happen?"

Take, for instance, the infamous story of Blockbuster turning down an opportunity to buy Netflix in the early 2000s. Blockbuster was the reigning king of video rentals, but Netflix was quietly growing into a digital powerhouse. Instead of seeing it as a future opportunity, Blockbuster's executives dismissed the idea, preferring to keep the old rental model, which eventually led to their downfall. Cutting off their nose to spite their face? Absolutely. They refused to adapt to a changing market and paid the price.

For product managers, avoiding "cutting off your nose to spite your face" is about making sure that decisions, even the tough ones, are made with the long-term goal in mind. It's about thinking through the impact of each move—not just reacting out of frustration or pride. It might mean letting go of a short-term win, like not arguing for an unnecessary feature, to ensure the overall success of the product. After all, sometimes the best way to succeed is to stop sabotaging your own potential—no matter how satisfying it might feel in the moment.

156. TEST THE WATERS

WHAT/WHEN

Verify before trusting.

> **"We decided to test the waters with a small-scale pilot before launching the new product nationwide."**

AS A PRODUCT MANAGER

When managing any large-scale operation, it's important to gauge new ideas or processes before fully committing. This is where the idiom "test the waters" comes into play—it's about trying something out on a small scale first to see if it's worth going all in. In the product management world, this could mean launching a new feature or policy in a limited capacity to understand how it might impact the broader system.

A great example of "testing the waters" in the retail industry was the rollout of Walmart's "Scan and Go" program. This innovative idea allowed customers to scan their items using an app and bypass traditional checkout lines. Before launching it in every store, Walmart tested the feature in select locations to see how customers responded and to work out any potential glitches. Once it proved successful, they expanded the program to more stores, refining the process along the way.

For a Product Manager, "testing the waters" often feels like standing at the edge of a pool, debating whether to dive in or just stick your toes in to check the temperature. You never know if the new process will be a splash or a flop, but that's the thrill of it. It's like running an experiment in real time: you're cautiously rolling out a new return process in a few regions to see if customers love it or if it leads to a chaos-filled customer service nightmare. And sometimes, it feels like you're testing water temperature blindfolded—but that's what keeps the job exciting!

157. WHITE ELEPHANT

WHAT/WHEN

An embarrassing condition or situation.

> "The new project was supposed to be a game-changer, but it quickly turned into a white elephant, with everyone pretending to care while secretly hoping someone else would take ownership."

AS A PRODUCT MANAGER

When it comes to group gifts, there's one tradition that always guarantees a few laughs, some confusion, and at least one person questioning their life choices: the "white elephant" gift exchange. This phrase perfectly captures the chaotic and often absurd nature of gift-giving when the goal is more about the spectacle than actual thoughtfulness. Whether it's a quirky knick-knack, an inexplicable item no one knows what to do with, or a surprise "treasure" no one ever asked for, the white elephant exchange is a time for everyone to laugh at the bizarre and unexpected.

A classic "white elephant" moment can be seen in the GitHub community during their annual "Hackathon" event, where developers are tasked with creating the most random, funny, or "useless" tool imaginable. One year, someone built a Chrome extension that replaced every instance of the word "bug" with "feature" in code reviews. It was a brilliant mix of absurdity and cleverness, much like a true white elephant gift. Everyone laughed, debated the utility of it (which was zero), and passed it around as though it were the best thing since sliced bread.

For Product Managers, participating in a "white elephant" exchange is like dealing with a bizarre GitHub pull request that, while entertaining, makes you question your life choices. Sometimes, you find yourself reviewing a feature request that is as absurd as that Chrome extension—something that might be hilarious to a niche group but has no actual purpose. But just like those random, offbeat projects on GitHub, sometimes it's the strange, unexpected ideas that provide the most amusement, even if they never actually make it to production.

158. ELEPHANT IN THE ROOM

WHAT/WHEN

This is an obvious issue that people avoid discussing.

> **"During the meeting, no one wanted to address the elephant in the room—our software is three months behind schedule, and the client's already asking for a refund."**

AS A PRODUCT MANAGER

When it comes to ignoring big issues in a room full of people, the "elephant in the room" idiom perfectly captures that moment when everyone is acutely aware of a problem, but no one wants to bring it up. This phrase is often used in situations where the obvious is being avoided, whether out of fear, discomfort, or denial. In business, it's like that one glaring financial shortfall or catastrophic PR crisis that everyone knows about but hopes will just fix itself somehow.

Take, for example, WeWork's infamous rise and fall. Before their failed IPO, the company was riding high with buzzwords like "community" and "disruption," but behind the scenes, there were some serious financial mismanagement issues and a questionable business model. No one in the room wanted to acknowledge the elephant—the fact that WeWork's valuation was based more on hype than solid fundamentals. The moment when the company's valuation tanked and the truth came out was the big reveal of the elephant everyone had been ignoring for months.

For Product Managers, dealing with the "elephant in the room" is like trying to pretend that the launch deadline for a new feature isn't in six hours, and you're still missing half the necessary components. Everyone sees it coming, but no one wants to point it out until it's completely unavoidable. It's like watching a slow-motion train wreck and hoping the train will somehow decide to stop before it hits the wall. But eventually, you have to acknowledge the elephant—after all, pretending it's not there just means you're going to have a lot more to clean up later. So, face it head-on, own it, and move forward, because ignoring it only makes the problem worse.

159. OUT OF THE BLUE

WHAT/WHEN

A surprise or an unexpected event.

> **"Out of the blue, our server crashed, and we found out the issue was a rogue line of code written by someone who definitely wasn't on the team anymore."**

AS A PRODUCT MANAGER

Sometimes, things just happen without warning, and you've got to be ready to react in an instant. The phrase "out of the blue" captures that sudden, unexpected nature of these situations. In the fast-paced world of marketing, this could mean having to pivot in response to a competitor's surprise move or jumping on an unexpected opportunity that appears without notice.

A prime example of a brand reacting "out of the blue" comes from Oreo during the 2013 Super Bowl. When a power outage caused a delay in the game, the Oreo social media team seized the opportunity to tweet a simple yet effective message: "You can still dunk in the dark." This spontaneous move was a hit, gaining widespread attention and making Oreo the hero of that Super Bowl, all thanks to a timely, unexpected response.

For marketers, reacting "out of the blue" is like being the one to notice that your product's moment has arrived just when you thought you were off the clock. It's about jumping on the opportunity without overthinking it, like when the competitor suddenly trips and falls, and you're right there with a perfectly timed ad. Of course, not every surprise will be a grand slam—sometimes, it's just a cake, half-baked and awkward—but every marketer knows that the thrill of that perfect, unexpected moment can make your brand feel alive and instantly relevant.

160. TAKE THE BULL BY THE HORNS

WHAT/WHEN

Aggressively taking charge of a situation.

> "We had a major bug in the system, but instead of waiting for a fix, I decided to take the bull by the horns and spent the weekend coding a workaround myself."

AS A PRODUCT MANAGER

When it comes to facing challenges head-on, some situations require a level of boldness that can only be described as "taking the bull by the horns." This idiom emphasizes taking immediate and decisive action, especially in tough or uncomfortable situations. Whether you're dealing with a high-pressure project or trying to fix a misstep before it spirals out of control, there's no room for hesitation.

Take, for instance, a real-world scenario like when Elon Musk decided to take charge and personally oversee the launch of SpaceX's Falcon Heavy rocket. Despite a history of setbacks and risks, Musk made the bold decision to push forward with the launch, even after previous failures. By literally taking the bull by the horns, he ensured the mission's success, making history with the rocket's successful launch and the iconic image of a Tesla roadster in orbit.

For Product Managers, "taking the bull by the horns" is all about owning up to a problem, whether you're facing a tough client, a tight deadline, or a product delay. It's not about waiting for someone else to swoop in and save the day; it's about diving in with both hands gripping that metaphorical bull's horns and steering the project in the right direction. Sometimes, that means saying no to a shiny new feature request because the current task at hand requires 100% of your attention. In the end, taking the bull by the horns is about embracing chaos, finding solutions, and getting things done— sometimes while you're being tossed around a bit. But hey, at least you'll have a great story for the post-mortem.

161. A DOLLAR SHORT AND A MINUTE LATE

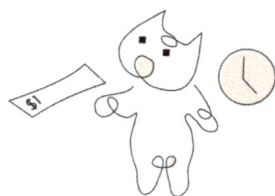

WHAT/WHEN

Failure due to bad timing or poor execution.

> "We were a dollar short and a minute late with the product launch, but at least the competitors were two dollars short and an hour late!"

AS A PRODUCT MANAGER

In business, timing is everything, and sometimes you're just a little too late to seize the opportunity. The phrase "A dollar short and a minute late" captures that feeling perfectly. It's that moment when you see the opportunity, but by the time you act, the window has closed. For companies, this could be missing out on a trend, a new market, or the chance to act before your competitors beat you to it.

A classic example of "A dollar short and a minute late" is Carvana's early years in the used car market. When Carvana launched in 2013, it had the idea of revolutionizing the car-buying experience by offering online sales and home delivery. However, they were a bit too early in the game. At the time, the market wasn't fully ready for online car buying, and traditional dealerships dominated the market. Carvana had to wait years until the industry caught up with their model. By the time the online shopping boom and COVID-19 pandemic hit, the market was finally ready, and Carvana's business skyrocketed. They had the right idea, but were just a bit too early to capitalize on it.

For Product Managers, "A dollar short and a minute late" is what it feels like when your innovative idea seems too far ahead of its time. You push for something groundbreaking, only to realize that the market or technology hasn't caught up yet, leaving you waiting for a shift in consumer behavior or infrastructure. It's like launching a feature that everyone would love... just a year too soon. But in the end, being ahead of the curve can still pay off—you just have to be patient, or in Carvana's case, get really good at waiting for the world to catch up.

162. MISS THE BOAT

WHAT/WHEN

Failure to exploit an opportunity.

> "We missed the boat on integrating that new AI agent, now our competitors are already sailing ahead with it!"

AS A PRODUCT MANAGER

In the world of product development, timing is crucial. If you wait too long, you might "miss the boat," meaning you've missed your opportunity. This idiom describes the feeling when you've hesitated or delayed making a decision, and now the window of opportunity has closed. In industries like AI and tech, where innovation happens at lightning speed, not acting quickly enough can leave you in the dust while competitors move ahead.

Take OpenAI, for example, when they released GPT-3 back in 2020. While other companies were still working on basic chatbots, OpenAI had already launched a cutting-edge language model that could generate human-like text across a variety of applications. Anyone who wasn't quick to notice or jump on the AI revolution at that time might've found themselves a bit behind, missing the boat as OpenAI's technology became a game-changer.

For product managers, "missing the boat" is that unsettling moment when the perfect solution or feature is right in front of you, but you hesitated for a second too long. It's like watching a rocket launch without you onboard, while you're still trying to figure out how to fill out the paperwork to get on the next one. You know it's not the end, but it sure feels like a missed opportunity when you're left watching others zoom past with the idea you could've had. It's a constant reminder that in fast-moving industries, sometimes you have to act before you even fully understand the journey ahead.

163. LEAVE IT ON THE FIELD

WHAT/WHEN

Make an all-out effort with nothing in reserve.

> **"We've got one last sprint before the product launch—let's leave it on the field and make sure everything's polished to perfection!"**

AS A PRODUCT MANAGER

When you're in the thick of things, it's essential to give it your all, no matter the situation. The phrase "leave it on the field" perfectly captures that mindset of putting everything you have into whatever you're doing, whether you're about to cross the finish line or fighting through the last few minutes of a game. In the world of business, this could be the moment when your team is putting the final touches on a project before a major deadline, with no energy left to spare.

Take the time Salesforce launched its groundbreaking cloud-based CRM system. The company was in full-on "leave it on the field" mode as they scrambled to meet the demands of a rapidly changing market and fierce competition. They pushed through late nights and endless iterations, making sure every feature was polished and every client need was met. The pressure was immense, but they gave everything they had, and the result was a product that revolutionized the industry and cemented Salesforce's position as a leader in the CRM space.

For product managers, "leaving it on the field" is about squeezing every last drop of effort into making sure a project succeeds. It's like preparing for launch and realizing you forgot to check one tiny but crucial detail—then jumping right back in and fixing it while time is running out. It's pushing through fatigue, the pressure, and the endless rounds of revisions, all while pretending you're not secretly wondering if your last email actually made sense (it didn't, but we'll fix it later). Sometimes, it's not about how well you did—just that you kept going until there was nothing left to give.

164. THE DIE IS CAST

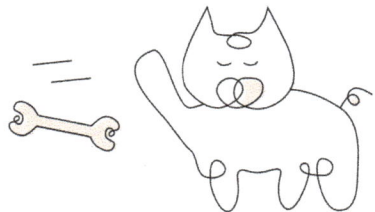

WHAT/WHEN

Decisions have consequences, good or bad.

> "After countless rounds of testing and feedback, we've made the final decision on the product design—looks like the die is cast, and we're rolling with it, for better or worse!"

AS A PRODUCT MANAGER

Sometimes, a decision hits you like a thunderbolt of inevitability. This idiom—"the die is cast"—perfectly captures those moments when you take action and realize there's no going back. It's like pressing the launch button and watching events unfold, whether you're ready or not. In the corporate world, this could mean greenlighting a risky product pivot or merging with a competitor despite last-minute jitters.

A prime example of "the die is cast" is when Sony decided to leap from the original PlayStation to the PlayStation 2—a massive bet on next-generation gaming. They poured millions into R&D and marketing, wagering that consumers would embrace DVD technology and cutting-edge graphics. Once Sony committed, there was no hitting the brakes, despite industry skeptics and steep manufacturing costs. The gamble paid off, forever cementing PlayStation as a titan in the gaming universe.

For product managers, "the die is cast" is that moment you finally lock your roadmap and say, "Ship it," while ignoring the butterflies in your stomach. It's the instant you realize there's no plan B—just a lot of crossed fingers and fervent Slack messages. And Chad, buddy, your brilliant idea for a last-second AR dancing giraffe feature? Still a hard pass. But hey, if everything goes sideways, at least you'll have a great story to tell over pizza at the next team retrospective.

165. THINK OUTSIDE THE BOX

WHAT/WHEN

Creative thought that conflicts with custom or convention.

> **"Great companies are built by leaders who think outside the box. They see the possibilities, not the limits."**

AS A PRODUCT MANAGER

Thinking outside the box is all about approaching a problem from a completely fresh angle. Instead of sticking to the same old methods, it's about breaking free from the constraints and finding new, creative solutions. In industries like design or product development, this idiom encourages innovation and the willingness to experiment with unconventional ideas. It's about solving problems in a way that no one else thought of—sometimes the best solutions are the ones that make you think, "Why didn't I think of that sooner?"

A real-world example of "thinking outside the box" is when T-Mobile challenged the traditional 2-year phone contract model. For years, customers were locked into long-term contracts that tied them down to one carrier, often with hefty penalties for breaking them. In 2013, T-Mobile introduced its "Un-carrier" strategy, eliminating 2-year contracts and allowing customers to pay for their phones in installments. This bold move not only disrupted the telecom industry but also forced other carriers to rethink their own business models. T-Mobile didn't just think outside the box—they threw the box out the window entirely.

For product managers, thinking outside the box is less about having a lightbulb moment and more about challenging the status quo. It's like the moment you realize that the way everyone else is doing things isn't going to get you the results you want—and deciding to try something completely off-the-wall. It's not just about creativity for the sake of creativity, though; it's about taking smart, calculated risks to push your product forward. Think of it like building a spaceship with only duct tape and rubber bands—and hoping it actually flies.

166. LEARNING CURVE

WHAT/WHEN

A rising proficiency due to exposure over a period.

> **"The learning curve on this new software is so steep, it's like trying to scale Mount Everest in flip-flops!"**

AS A PRODUCT MANAGER

It's essential to be prepared for challenges that come with new experiences, especially when they seem overwhelming. The phrase "learning curve" perfectly captures the idea of stepping into the unknown, where growth and progress are built by overcoming initial obstacles. In the world of product management, it's the metaphorical hill you climb when starting a new project, tool, or process that feels like a mountain—only to realize you're just learning to hike.

Take for example a product manager switching from a complex project management tool like Jira to the more intuitive monday.com. The learning curve is dramatically different, with monday.com's simple, user-friendly interface making it much easier to get started. While Jira might require extensive setup and a deep dive into its intricate workflows, monday.com offers a more straightforward approach, with drag-and-drop functionality and easy-to-understand templates that allow a new user to hit the ground running. The transition is much less about trial and error and more about quickly discovering how best to use the platform to organize tasks and collaborate effectively.

For Product Managers, the "learning curve" is more about getting comfortable with discomfort than anything else. It's like being thrown into the deep end of a pool you're not sure you know how to swim in, only to realize the pool has 17 different depths and a slide that's always just out of reach. And while you might be floundering for a bit (cue the drowned laptop in your bag), eventually you start swimming, and the panic shifts to "okay, I'm not perfect yet, but I'm definitely not drowning anymore." At the end of the day, the "learning curve" might make you feel like you've just tried to solve a Rubik's cube blindfolded, but once you crack it, that satisfaction feels pretty darn good.

167. BY THE BOOK

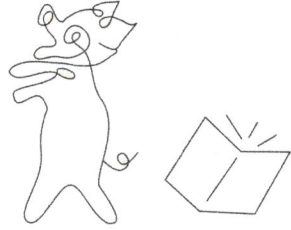

WHAT/WHEN

Complying with rules, regulations, and accepted practices.

> **"Noah's disadvantage is his lack of creativity. He does everything by the book."**

AS A PRODUCT MANAGER

It can be tempting to cut corners sometimes, especially when deadlines loom overhead. This idiom perfectly embodies the importance of following established procedures from start to finish. In the tech industry, "by the book" can mean anything from thoroughly documenting APIs to meticulously triple-checking release notes before clicking "Deploy."

One real-world example of a product manager going "by the book" is when NASA oversaw software development for the Mars Rover missions. Each step had to be executed according to meticulously crafted protocols because a simple oversight on Earth could become a catastrophe on another planet. The product manager adhered so strictly to the rulebook that even a single line of code went through multiple reviews before it beamed off to the Red Planet.

For Product Managers, "by the book" is less about being a buzzkill and more about making sure the entire project doesn't burst into chaos just because someone felt like freestyling the compliance checklist. It's juggling a hundred constraints with a calm demeanor—like confirming every last bit of data security while the marketing team begs to rename half the features two days before launch. Sometimes, it feels like you're simultaneously the world's most diligent librarian, the code police, and the person who alphabetizes their cereal shelf. But hey, nothing cements team unity like that collective sigh of relief when everything passes audit without setting off a single alarm.

168. CUT CORNERS

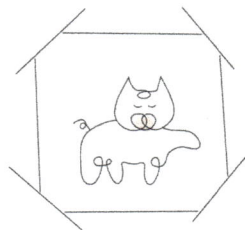

WHAT/WHEN

Reconfiguring a process to save time, money, and resources.

> **"Our development team tried to cut corners on the app's security, but now we're all just hoping the hackers don't notice."**

AS A PRODUCT MANAGER

When it comes to getting things done, sometimes people take shortcuts. This idiom captures the temptation to "cut corners" in order to save time or effort, though it often comes at the expense of quality. In a business setting, this can show up in decisions that prioritize speed over thoroughness—like rushing a product launch without proper testing or skipping key steps in a process to meet a deadline.

A real-world example of cutting corners occurred in 2023 when the airline Southwest faced massive flight cancellations due to a software glitch. The glitch was a result of hasty updates to their scheduling systems that weren't fully tested, and it ended up grounding thousands of flights and causing widespread chaos. In this case, the decision to cut corners on software testing led to not just an operational nightmare, but a significant blow to customer trust and loyalty.

For product managers, cutting corners can feel like a delicate balance between making progress and praying your decisions don't come back to bite you. It's all about deciding whether you can take a shortcut or if you need to go the extra mile—like the time you pushed out a feature without full testing because marketing insisted it was a "top priority" (spoiler: it wasn't). Sure, everyone loves a fast solution, but when it backfires, it's like trying to get out of a speeding ticket with a joke about "well, at least I was driving fast." It's a gamble, and not one you'd want to make every day.

169. CHOP-CHOP

WHAT/WHEN

Directions to hurry, i.e., move quickly.

> "Alright team, we've got a tight deadline, so let's chop-chop and get those product specs finalized before lunch!"

AS A PRODUCT MANAGER

When someone says "chop-chop," they're really calling for some quick action, no dilly-dallying. It's all about getting things done at lightning speed, especially when time is of the essence. Whether you're in a rush to meet a deadline or trying to wrap up a task that's been hanging over your head, the phrase "chop-chop" brings the urgency to the table, reminding you to step it up, now.

A memorable example of "chop-chop" in action comes from the 2019 Cybertruck demo. When Tesla unveiled the futuristic vehicle, they intended to show off its indestructible armored glass windows. Instead, the demo turned into a bit of a disaster when the window shattered on stage after a quick toss of a metal ball. The whole thing could have been a major setback, but Elon Musk and his team didn't waste any time—chop-chop, they pivoted and made jokes about it on the spot, turning the mishap into a humorous moment rather than a full-blown crisis.

For product managers, "chop-chop" is less about frantic panic and more about keeping momentum when things go wrong. It's about turning on a dime and finding solutions before the clock runs out. Like dealing with a major bug just before a product launch while someone in the background asks if they can add one more tiny feature (hint: they can't). It's fast thinking, fast talking, and a whole lot of prioritizing, even when it feels like you're juggling flaming torches. Sometimes it's all about moving quickly and fixing things on the fly, hoping your team doesn't notice the occasional crack in the windshield.

170. GREASE THE WHEELS

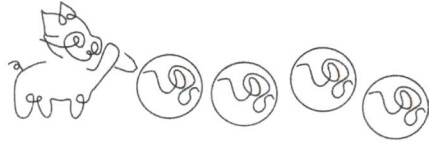

WHAT/WHEN

Minimize obstacles that might arise in the pursuit of an objective.

> "I was worried we wouldn't get permission to use the location. We're lucky Barry knew the owners and greased the wheels to get agreement."

AS A PRODUCT MANAGER

In the world of big data, sometimes things don't always go as planned, and the solution often requires a little extra effort to make sure everything runs smoothly. The idiom "grease the wheels" is a perfect fit for this kind of situation. It's about ensuring that everything functions without friction, whether it's facilitating communication between departments or resolving logistical issues that could delay progress.

Take Snowflake, for example. Imagine a scenario where multiple teams are working on a data migration project, and the product manager is tasked with coordinating the efforts. There are integrations to consider, deadlines to meet, and countless moving parts. The manager's job is to "grease the wheels" by streamlining processes, clearing up miscommunications, and making sure each team has what they need to stay on track. It's about smoothing out the roadblocks before they cause serious delays.

For Product Managers working with Snowflake, "greasing the wheels" isn't about throwing a wrench in the works but about keeping things running without hiccups. It's managing the behind-the-scenes details so teams can stay focused on their main tasks. Whether it's ensuring the database migrations go off without a hitch or making sure the right stakeholders are looped in on time-sensitive updates, you're the one keeping the gears turning. And though you might not always get credit for the smoothness, the end result is a product launch that's as seamless as a well-oiled machine—until, of course, the next unexpected bug rolls in.

171. NUTS AND BOLTS

WHAT/WHEN

The many details and activities that are critical to achieving a goal.

> "We were stuck in endless meetings, but once we got to the nuts and bolts of the project, we finally made progress."

AS A PRODUCT MANAGER

When it comes to running a smooth operation, understanding the "nuts and bolts" of a system is key. This idiom highlights the essential, often overlooked details that keep everything running like clockwork. In business, it can refer to the core processes or tools that are necessary to make things work behind the scenes—like the gears that turn in the engine of a car, or in this case, the system you're managing.

A great example of a product manager diving into the "nuts and bolts" of things can be seen in the launch of the first iPhone. Steve Jobs didn't just rely on the flashy new touchscreen and the sleek design; there were countless behind-the-scenes technical challenges to ensure that everything—from battery life to app compatibility—worked as expected. Product managers had to be deeply involved in the nitty-gritty, ensuring that each component fit together seamlessly and worked properly from day one.

For Product Managers, understanding the "nuts and bolts" is more than just keeping the gears greased and turning; it's about not letting the whole operation fall apart because one screw was loose. It's like trying to fix a blender while it's still running and hoping you don't end up wearing your morning smoothie. Managing a product is about recognizing where the tiny problems are hiding, and fixing them before they spiral into bigger disasters. It's a delicate dance between technical troubleshooting and strategic decisions—because in the end, if one part breaks, the whole thing might come crashing down. But hey, nothing builds team camaraderie like a good session of unscrewing the tangled mess that nobody saw coming.

172. RUN INTO A BUZZSAW

WHAT/WHEN

Encountering an obstacle with severe consequences.

> **"We were all set to launch the new app, but we ran into a buzzsaw with the last-minute security audit – looks like we'll be debugging till the end of the year!"**

AS A PRODUCT MANAGER

Sometimes, you think everything's going smoothly, and then, out of nowhere, you run into a buzzsaw. This idiom perfectly describes those situations when you're charging ahead, only to be suddenly stopped by an obstacle so big and unavoidable that it feels like you're out of options. Whether it's a competitor with a superior product or a technical failure you couldn't predict, the buzzsaw is that moment when things go sideways, and you're left wondering how to recover.

A notable example of Square "running into a buzzsaw" happened in 2015 when the company launched their mobile credit card reader for the iPhone 6. They anticipated smooth sailing, but Apple's new phone came with changes that made Square's reader incompatible with the updated iPhone. The timing was disastrous for Square, as they were positioning their product as the go-to solution for small businesses. They had to scramble, redesign the product, and quickly release a new version, all while managing the public's frustration.

For Product Managers, "running into a buzzsaw" means realizing that even with meticulous planning, things can still go terribly wrong. It's when your sleek, carefully crafted product hits a roadblock so major that it feels like your whole strategy is being torn apart. At that moment, it's all about managing the chaos, patching things up on the fly, and keeping your cool while knowing that you've got a long road ahead to get back on track. And let's face it— there's no better bonding moment for a team than collectively tackling a buzzsaw-sized disaster with a blend of panic, humor, and the determination to fix it.

173. BLOW A FUSE

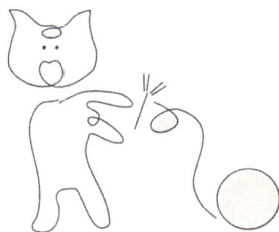

WHAT/WHEN

Lose one's temper in an exaggerated manner.

> **"During the team meeting, Dave totally blew a fuse when the server crashed right before the big product demo."**

AS A PRODUCT MANAGER

When you're working in high-pressure environments, it's crucial to maintain control—especially when things go wrong. The idiom "blow a fuse" is an apt way to describe that exact moment when everything gets too overwhelming and your patience just can't take it anymore. This idiom is often used when someone loses their temper or becomes visibly frustrated, which can sometimes happen in even the most controlled situations.

Take, for example, a real-world case from Coinbase during the 2021 crypto surge. As the price of Bitcoin soared, Coinbase's platform experienced major outages, frustrating thousands of users who couldn't access their accounts at a crucial moment. The customer support team, already overwhelmed with inquiries, saw a significant increase in angry complaints. It's not hard to imagine how someone, after handling countless irate emails and Twitter rants, could blow a fuse—especially when dealing with clients who want answers immediately but only get an automated response.

For Product Managers, "blowing a fuse" is less about losing your cool and more about keeping it together despite the constant barrage of demands. It's like trying to assemble a jigsaw puzzle with missing pieces while a toddler insists they need your full attention. Sure, you might snap when your team member asks why the meeting is still going on after 90 minutes—when clearly, we're all out of answers. But hey, at least you didn't throw your laptop out the window (this time). Sometimes, blowing a fuse is just the temporary glitch before getting back to calmly putting out fires—and hopefully, with fewer sparks.

174. MURPHY'S LAW

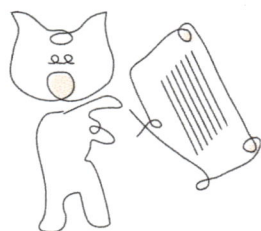

WHAT/WHEN

What can go wrong will grow wrong.

> **"Of course, the server went down right after we launched the new feature—guess Murphy's Law is working overtime today!"**

AS A PRODUCT MANAGER

Things rarely go as planned, and that's where Murphy's Law swoops in. If something can go wrong, it will. This idiom is a humorous reminder that life is full of unexpected turns, and sometimes the universe just seems to have it out for you. In the world of tech, this can mean anything from a server crash in the middle of a big presentation to your Wi-Fi cutting out when you're about to submit a critical report.

A recent example of Murphy's Law in action was in 2020 when Red Hat faced major issues with its enterprise Kubernetes offerings. Despite the massive investment in automation and testing, the rollout of certain updates led to unanticipated compatibility issues with users' existing setups, resulting in widespread system outages. The fallout was frustrating for both customers and internal teams, as something that was supposed to streamline cloud-native deployments turned into a headache, proving that even the most trusted brands can fall victim to Murphy's Law.

For Product Managers, Murphy's Law is like trying to keep all the plates spinning while one of them is definitely cracking, and there's always that one guy—looking at you, Chad—who still thinks adding "one more feature" will somehow make everything better. It's that constant reminder that no matter how much you plan, the universe will throw a wrench in your well-oiled machine, and you'll be left juggling broken code, upset customers, and an unhelpful inbox. But hey, if nothing ever went wrong, you wouldn't have those priceless stories for happy hour.

175. ALL SOUND AND FURY

WHAT/WHEN

All promotion with little substance.

> **"After weeks of endless meetings and no clear direction, the product launch felt like all sound and fury, signifying nothing."**

AS A PRODUCT MANAGER

It's easy to get caught up in the noise, but sometimes you have to ask yourself: "Is this a real issue, or is it just all sound and fury?" This idiom captures the idea of something that creates a lot of commotion or drama but ultimately lacks substance or significance. In the world of business and tech, this is particularly relevant when a lot of people are making a big deal about minor problems that don't actually impact the bottom line.

A good example of "all sound and fury" in action occurred during the 2016 launch of the Samsung Galaxy Note 7, which had some highly publicized battery issues that led to fires. Amid the hysteria, there were tons of rumors and wild stories circulating, but the core issue was really just a faulty battery design. While it was a serious issue for some, much of the commotion was just loud, dramatic noise fueled by social media and news outlets.

For Product Managers, "all sound and fury" is a reminder to separate the meaningful issues from the distracting ones. It's easy to get caught up in the drama when someone is passionately arguing over a feature that barely anyone uses (looking at you, Greg from QA). But sometimes, you have to take a step back and ask, "Is this truly a crisis, or is it just a lot of yelling and keyboard pounding?" It's about making sure you're addressing real problems without getting sidetracked by the noise. After all, managing a product is less about making a lot of noise and more about making meaningful progress.

176. BLITZ

A surprise, substantial effort for immediate impact.

> **"We're going to blitz through this product launch so fast, even our competitors won't know what hit them— hopefully, in a good way!"**

AS A PRODUCT MANAGER

When things need to be handled at lightning speed, there's no better phrase than "blitz." This idiom is perfect for those moments when a full-on, no-holds-barred approach is needed to tackle something quickly and with maximum effort. In the world of product management, it could mean rallying the team to launch a product or fix an urgent issue all within a crazy short time frame.

A prime example of a "blitz" in action happened when Amazon Prime Day first launched. The company had one goal: sell as much stuff as possible in just 48 hours. The product managers had to make sure every aspect of the event—from site stability to stock levels—was dialed in and running without a hitch. With all hands on deck, they blitzed through any issues, sometimes in real-time, ensuring that the world's biggest online sale didn't collapse under the weight of its own success.

For Product Managers, a "blitz" often feels like having to sprint through an obstacle course of issues while blindfolded. It's the pressure of delivering a project on a razor-thin deadline while your coffee is getting cold, and you're pretty sure someone on the team just sent you a Slack message about a new "critical" feature request (spoiler: it's not). It's all about making things happen—fast—while somehow still maintaining your cool. Picture it as a sprint, but with a few hurdles, a ticking clock, and an occasional tumble into a pit of bugs. It's chaotic, it's intense, and it's a whole lot of fun when it's over.

177. THE SHORT AND CURLIES

WHAT/WHEN

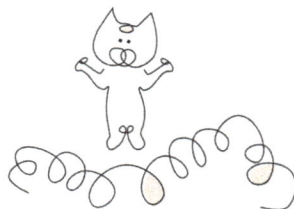

A description of a critical situation.

> "After the investor meeting, the CTO really laid out the short and curlies of the new product roadmap—no fluff, just straight to the tough decisions."

AS A PRODUCT MANAGER

Sometimes life has a funny way of pinning us in tight corners, and that's precisely what this phrase highlights. "The short and curlies" succinctly captures that moment when you're well and truly stuck, with no elegant way to wriggle free. In the tech world, it could be anything from discovering that your entire codebase is out-of-date right before launch to realizing you signed a vendor contract that includes a non-negotiable, last-minute upgrade fee—ouch.

One real-world example of a product manager "caught by the short and curlies" is when Samsung had to deal with the Galaxy Note 7 battery fiasco in 2016. The phones had a tendency to overheat and, in some cases, burst into flames—definitely a situation with zero room for error. The product manager had to navigate between media scrutiny, customer outrage, and the race to identify a fix, all while attempting to preserve Samsung's reputation (and its customers' pockets and pants).

For Product Managers, being "caught by the short and curlies" isn't about bravado; it's about trying not to topple the whole Jenga tower while you figure out which block started this mess in the first place. It's juggling last-minute changes with a straight face—like someone on the team suggesting a total UI overhaul two days before launch, even as you're on a video call explaining to the CEO why half of Europe can't log into the app. Sometimes, it feels like you're both the hostage and the hostage negotiator. But hey, nothing says "team bonding" like collectively sweating bullets over a release that was supposed to go smoothly—famous last words.

178. BETWEEN A ROCK AND A HARD PLACE

WHAT/WHEN

A choice between two equally bad solutions.

> **"I'm between a rock and a hard place trying to choose between upgrading the software or fixing the server — both will cost, but only one will get me a lunch break!"**

AS A PRODUCT MANAGER

It's not unusual to find yourself stuck between a difficult choice and an even worse one. This idiom really nails that feeling of being trapped in a no-win situation, where you're forced to make a decision but none of the options feel right. Whether it's about picking the lesser of two evils or navigating a sticky business dilemma, "between a rock and a hard place" speaks to that feeling of impending doom when it feels like every move could make things worse.

One of the most recent examples of being "between a rock and a hard place" is the FTX debacle. Sam Bankman-Fried's cryptocurrency exchange crashed spectacularly in 2022, dragging down billions of dollars in investments and tarnishing the reputations of several high-profile endorsers. Celebrities like Tom Brady, Gisele Bündchen, and Larry David had all publicly endorsed the platform, which ended up being a massive Ponzi scheme. They found themselves stuck in a PR nightmare, caught between defending their own involvement and distancing themselves from a scandal that was unfolding in real-time. No matter how they tried to spin it, the damage was already done.

For Product Managers, being "between a rock and a hard place" is a lot like trying to make sense of a disastrous PR crisis while keeping the lights on. It's balancing the pressure of keeping stakeholders happy while the product you're launching starts to unravel under the weight of conflicting priorities. You're the one who has to explain why the latest feature is delayed (again) while quietly wondering if anyone will still trust your product after all the damage control. It's kind of like being caught in a reality show where no matter how you play the game, you're probably getting voted off the island anyway. But hey, at least you get to wear your survival skills like a badge of honor.

179. PUT YOUR EGGS IN A SINGLE BASKET

WHAT/WHEN

Risking everything on a single element.

> "Spending our whole ad budget on sponsored YouTube vids seems risky. We're putting our eggs in a single basket."

AS A PRODUCT MANAGER

When it comes to making strategic decisions, focusing all your attention on one path can be a risky move. The idiom "don't put all your eggs in one basket" perfectly encapsulates the need to diversify your efforts. Whether it's a product, a market, or a technology, betting everything on one thing can lead to disastrous results if that single focus doesn't pan out.

Take Intel, for example. For years, Intel dominated the PC chip market, focusing exclusively on processors for personal computers. While they were king of the PC world, they completely overlooked the growing mobile market. As smartphones started to take off, Intel's focus on PCs left them scrambling to catch up, allowing competitors like ARM to thrive in mobile processors. Had Intel diversified their approach earlier, they might have been able to secure a stronger position in the mobile industry before it exploded.

For Product Managers, Intel's story is a cautionary tale about the risks of "putting all your eggs in one basket." It's easy to get comfortable with the success of a single product or market, but diversification is the key to long-term survival. You might not have the resources to conquer every new trend, but keeping an eye on emerging markets and technologies can prevent your product from becoming obsolete. Just like Intel learned the hard way, relying too heavily on one "basket" can leave you scrambling when the landscape changes.

180. NO STONE UNTURNED

WHAT/WHEN

A thorough effort, including options.

> **"We're leaving no stone unturned in our quest to find the next big thing in AI, even if that means asking the office plants for ideas."**

AS A PRODUCT MANAGER

When tackling complex tasks, sometimes you just have to be thorough—like a detective sifting through every clue to solve a mystery. The idiom "no stone unturned" emphasizes the importance of leaving no possibility unexplored in order to find the solution. Whether it's combing through mountains of data or carefully considering every potential avenue, this approach is key to finding the best answers.

Take the example of NASA's Mars Rover mission in 2021, when the team worked tirelessly to address a software malfunction. The engineers didn't just fix the immediate problem—they went over every single line of code and piece of hardware, making sure they left no stone unturned in order to ensure the Rover would keep running smoothly. In this case, "no stone unturned" meant looking at every possibility, from software bugs to physical wear and tear on the equipment.

For product managers, "leaving no stone unturned" often involves diving into the weeds, getting elbow-deep in details, and checking every single thing, even when it feels like you've already checked it a hundred times. It's like going through your app's user flow for the 10th time because you think you missed a typo or double-checking that the new feature doesn't crash the entire platform (it's not "just a bug," Chad). Sometimes it's a game of detective work, where the solution might be hidden in the most unexpected places, like the very last line of code or an email from three months ago that *might* have some critical insight. But hey, nothing says "dedication" like finding that one elusive issue hiding under a pile of code, documentation, and caffeine-induced paranoia.

BONUS

181. LET SLEEPING DOGS LIE

WHAT/WHEN

Don't encourage problems that can be avoided.

> "Let's just let sleeping dogs lie and not bring up that outdated software issue during the meeting—nobody wants to wake that beast."

AS A PRODUCT MANAGER

In the fast-paced world of digital banking, there's often a temptation to dive into every small issue as soon as it arises. However, the phrase "let sleeping dogs lie" is a reminder that sometimes it's smarter to avoid stirring things up when everything is functioning smoothly. In the context of neo-banks, this could mean resisting the urge to tweak a feature that's working just fine, especially when it could potentially disrupt the user experience.

Take, for example, the rise of Chime, a popular neo-bank that made a name for itself by offering simple, no-fee banking. In the early stages, the company could have added all sorts of complex features to differentiate itself from traditional banks. But instead, they wisely chose to stick to their core offering of easy-to-use, no-frills services that customers loved. By letting the "sleeping dog" of complexity lie, they were able to focus on building a loyal customer base without unnecessarily complicating things.

For product managers in the neo-bank space, "letting sleeping dogs lie" is about recognizing when to resist overcomplicating the product. Sometimes it's tempting to add a bunch of bells and whistles, like a new rewards program or a flashy interface update. But often, the best move is to leave well enough alone—just like when your app is running smoothly and a customer suggests adding a feature that might not add value. The key is knowing when to let things be and when to stir the pot. In the digital banking world, doing nothing can sometimes be the smartest thing you do.

182. BEAT A DEAD HORSE

WHAT/WHEN

Continue to cycle a closed subject.

> "There's no need to beat a dead horse by revisiting the same issue in every meeting; let's focus on finding solutions instead."

AS A PRODUCT MANAGER

In product management, it's important to recognize when you're circling around a point that's already been resolved or is no longer relevant. The idiom "beat a dead horse" captures this perfectly—it's about not wasting energy on something that can't be changed or should just be left behind. In the tech world, especially with blockchain technology, this phrase often comes into play when teams keep rehashing a failed idea or old technology that's no longer going to make an impact.

A good example of "beating a dead horse" in the blockchain world is the early hype around Initial Coin Offerings (ICOs). In 2017, ICOs were the hottest thing since sliced bread, with companies launching tokens to raise millions of dollars. However, many projects didn't deliver on their promises, and regulators started cracking down. Yet, some teams continued to try and revive their ICO strategies, holding endless discussions and hoping for another breakout moment despite clear signs the market had moved on. The reality was that it was time to pivot, but some couldn't resist continuing to flog that tired horse.

For Product Managers working with emerging technologies like blockchain, "beating a dead horse" is about recognizing when it's time to stop rehashing old ideas or pursuing a path that's no longer viable. It's not about abandoning innovation, but knowing when a once-promising technology or strategy is no longer worth investing time or resources into. Like those teams still stuck on ICOs when everyone else is focusing on more sustainable blockchain applications, it's crucial to redirect your energy toward the next big thing— before you end up stuck in a loop with a horse that's already been turned into glue.

183. HARD TO SWALLOW

WHAT/WHEN

Unbelievable.

AS A PRODUCT MANAGER

When something is "hard to swallow," it's a polite way of saying you've just encountered a truth or situation that's not exactly easy to digest. This idiom often pops up when you're faced with unpleasant facts or uncomfortable situations that make you want to spit them out rather than accept them. In the corporate world, this could mean having to deal with feedback that's less than stellar or realizing that a project you've poured hours into just isn't going to work.

A classic real-world example of something "hard to swallow" is when MySpace was offered the chance to buy Facebook in its early days. Mark Zuckerberg, the founder of Facebook, approached MySpace with a proposition to sell his growing social media platform for just a few million dollars. MySpace, at the time the king of social networking, turned it down, thinking that Facebook wouldn't pose a real threat. Fast forward a few years, and that decision was a bitter pill to swallow as Facebook skyrocketed in popularity while MySpace faded into irrelevance.

For Product Managers, "hard to swallow" often means dealing with harsh realities that no one wants to face. It's like being told your product idea is a complete flop after you've already spent months hyping it up in meetings (yep, Chad, that one). Sometimes, it's realizing that the budget you've been working with is about to get slashed—just as you're about to launch. But no matter how tough it is, it's all about taking that information in stride, pretending like you're not choking on the truth, and figuring out what to do next—without the team noticing your internal gag reflex.

184. SAUCERED AND BLOWED

WHAT/WHEN

Humorously describes a situation where things have gone disastrously wrong.

> **"The new software update saucered and blowed the entire system, leaving us scrambling to fix bugs before the client demo."**

AS A PRODUCT MANAGER

When things go sideways in a way that completely catches you off guard, it's time to embrace the chaos and maybe even chuckle a little. The idiom "saucered and blowed" perfectly describes the shock and disbelief you feel when something unexpected and wildly disruptive happens. It's like that moment when you've just poured yourself a cup of coffee, and suddenly your saucer flies off the table in shock. In the business world, "saucered and blowed" encapsulates the feeling of being blindsided by a situation you never saw coming.

A prime example comes from SoftBank's investment in WeWork. The company had invested heavily in the shared office space startup, believing it was the next big thing. However, when WeWork's IPO plans fell apart in 2019, SoftBank was left scrambling. The company's valuation plummeted, and SoftBank's massive investment turned into a costly disaster. They were "saucered and blowed" by the rapid unraveling of a once-promising investment, and suddenly, the once-glamorous future they had envisioned looked much more like a cautionary tale.

For product managers, a "saucered and blowed" moment can happen when a deal you thought was secure crumbles at the last minute or when an unexpected crisis turns your carefully crafted roadmap upside down. It's a situation where you have to recalibrate and rethink everything on the fly, hoping you can somehow salvage the situation. But just like SoftBank, you may find yourself sitting there, wide-eyed, wondering how everything could fall apart so quickly. In those moments, it's important to remember that even the biggest players get caught off guard—so take a breath, reassess, and be ready to pivot. After all, every setback is a chance to make a comeback.

185. SCREEN DOOR IN A SUBMARINE

WHAT/WHEN

A description of futility or ineffectiveness.

> "Trying to scale our cloud infrastructure with outdated servers is like putting a screendoor in a submarine—it's not going to hold up under pressure."

AS A PRODUCT MANAGER

When it comes to launching a product, the devil is in the details. This idiom perfectly illustrates the idea that no matter how promising the concept or the marketing may be, if the foundational aspects are flawed, the whole thing can fall apart. It's like trying to sell a car that looks great but has no engine. In the business world, this highlights the importance of making sure everything is in place before you push the pedal to the metal.

A classic example of the "screendoor in a submarine" mentality was Jawbone, a company that once made waves in the wearables market with its fitness trackers and Bluetooth headsets. Despite early success and a sleek design, Jawbone's products suffered from serious reliability issues. The devices were often prone to malfunctions, and customer complaints about the poor battery life and syncing problems piled up. No matter how many times Jawbone tried to rebrand or release new versions, the core issues couldn't be solved, and the company eventually folded. They were trying to sell a luxury product that simply didn't have the reliability to back it up.

For Product Managers, the "screendoor in a submarine" lesson is clear: don't ignore the basics, no matter how shiny your product might seem. It's tempting to get caught up in the buzz, but if the core functionality isn't there, you're just patching leaks with a bad fix. Much like Jawbone's struggle, you can only get so far before the cracks in the foundation start to show. In those situations, it's not about salvaging what you can; it's about knowing when to rethink the whole approach before the ship sinks.

186. CHEW THE FAT

WHAT/WHEN

Engaging in a long, cordial conversation, usually in an effort to put participants at ease.

> "Before we dive into the quarterly numbers, let's just chew the fat for a minute and see how everyone's feeling about the new software rollout."

AS A PRODUCT MANAGER

When you're working on a project or navigating a tricky situation, it's essential to carve out time for some casual conversation—especially when it helps build rapport or ease tension. The phrase "chew the fat" speaks to the kind of laid-back, no-pressure discussions that keep things moving smoothly. It's not about jumping into business right away; it's about letting things breathe and bonding before tackling the serious stuff.

A great example of "chewing the fat" in action happens in meetings at companies like Google, where employees often take a few moments to chat about everything from new tech trends to last night's dinner before diving into project updates. These conversations help build stronger relationships and set a collaborative tone, allowing everyone to feel more comfortable before jumping into the grind of work. It's the human version of "warming up" before a marathon, except no one's running, and the only sweat is from the coffee machine that's perpetually broken.

For product managers, "chewing the fat" isn't about avoiding work—it's about keeping the team's morale high and ensuring communication stays fluid. It's those casual chats in the breakroom about weird dreams, weekend plans, or office gossip that can spark creativity or uncover hidden issues before they become full-blown problems. Sometimes, you might find that your best product idea was born from a conversation about the most bizarre sandwich someone's ever eaten. While it may seem unproductive, it's like sharpening your tools before tackling the wood—sometimes, you need a moment to chew the fat before you start hammering away at the real work.

187. ONE-TRACK MIND

WHAT/WHEN

Focused on a single idea, task, or goal to the exclusion of all others.

> **"His one-track mind kept him focused on optimizing the algorithm, even when the rest of the team was brainstorming ways to scale the entire platform."**

AS A PRODUCT MANAGER

It's easy to get distracted by a thousand different things, but some people just can't seem to think about anything except the task at hand. The idiom "one-track mind" perfectly captures this ability—or lack thereof—to focus solely on one thing, often to the exclusion of everything else. In the fast-paced world of product management, it can be both a blessing and a curse when you find yourself hyper-focused on a single project, ignoring all other priorities.

Take, for example, Jeff Bezos in the early days of Amazon. In the 90s, when the company was struggling to get off the ground, he had a relentless focus on growth. His one-track mind was fixated on making Amazon the go-to marketplace for books, and later, for everything else. While other investors and executives were worried about short-term profits, Bezos stayed single-minded, thinking about long-term customer obsession. His focus paid off, and Amazon is now the e-commerce behemoth we know today.

For product managers, having a "one-track mind" is less about genius focus and more about missing the forest for the trees. It's like diving so deep into a user interface design that you forget to check whether the app actually works. Sometimes, you find yourself so consumed by perfecting a feature that you can't even remember what the original project goal was—until someone asks, "Wait, when are we going to launch this?" Oh right, that little detail! It's a reminder that while focusing on one thing can be productive, it's just as important to zoom out and see the bigger picture before your single-track turns into a one-way street to burnout.

188. BLANK CHECK

WHAT/WHEN

Complete freedom of action with no limits or boundaries.

> **"After securing the funding, the startup essentially had a blank check to scale, though the CEO still insisted on 'not buying a yacht just yet.'"**

AS A PRODUCT MANAGER

In the fast-paced world of business, sometimes you need to give someone the "blank check" to get things done. This idiom captures the idea of offering full trust and freedom to take action without limits or second-guessing. In a corporate setting, it can happen when a manager gives a team member carte blanche to innovate or solve a problem, fully trusting them to make the right call without micromanagement.

A perfect real-world example of a "blank check" moment can be seen in the world of AI. Many AI scientists and startups are getting funded with what feels like a literal blank check. Investors, eager to cash in on the next big breakthrough, are pouring millions into AI ventures, often without fully understanding the technology or its future implications. Just a few years ago, a small AI startup might have struggled to secure funding, but today, they can get a blank check simply by mentioning machine learning or deep learning in their pitch. The rapid growth and hype around AI have turned it into a business space where trust and potential seem unlimited, even if the results are still up in the air.

For Product Managers, handing out a "blank check" is less about writing a literal check and more about empowering your team to do their best work. It's about offering them the trust and support to think outside the box—no questions asked. This doesn't mean anything goes; it's more like saying, "I trust you to make this work, but please don't set anything on fire." Because let's be real: giving a blank check isn't always risk-free, but it does foster a team environment where bold decisions are the norm, and the results are often worth the risk.

NOW IT'S YOUR TURN TO MAKE PIGS FLY!

Congratulations! You've just flipped through 180 idioms and phrases that are now in your toolkit to elevate your communication and storytelling as a product manager. But let's face it—language is as dynamic as your last product roadmap (yes, the one that went out the window). There's always room for more!

Did I miss an idiom that has helped you rally a team or win over stakeholders?

Do you have a phrase that's so good it deserves its own slide deck? Well, don't keep it to yourself! Head over to **www.ali.ink** and share your gems.

Not only will you get bragging rights if your idiom makes it into the second edition of Make Pigs Fly, but you'll also be immortalized in print as a contributor. That's right—your name, in a book, next to phrases like "skin in the game" and "move the needle." It doesn't get more legendary than that.

So go on, let your creativity fly. Or, as I might say, don't let the cat get your tongue. See you on the site—and maybe in the next edition!

NOTES

www.ingramcontent.com/pod-product-compliance
Lightning Source LLC
Chambersburg PA
CBHW041208220326
41597CB00030BA/5096

* 9 7 9 8 9 9 2 5 4 1 6 3 2 *